JOEL SPENCER
Courant Institute
New York University

Ten Lectures on the Probabilistic Method
Second Edition

SOCIETY FOR INDUSTRIAL AND APPLIED MATHEMATICS

PHILADELPHIA, PENNSYLVANIA 1994

Printed by Capital City Press, Montpelier, Vermont

Contents

Preface to the Second Edition

The first edition of this monograph was compiled from the notes for a series of ten lectures given at the CBMS-NSF Conference on Probabilistic Methods in Combinatorics held at Fort Lewis College in Durango, Colorado in 1986. The Durango Lectures were a Rocky Mountain high. The writing, done immediately after the lectures, was breezy and effortless. I attempted to remain close to the content, order and spirit of the lectures. I retained the more informal, first-person, sometimes rambling narrative style of the lectures.

The "probabilistic method" is a powerful tool in graph theory and combinatorics. Explained very basically, this method proves the existence of a configuration, first, by creating a probability space whose points are configurations, and second, by showing a positive probability that the "random configuration" meets the desired criteria. As the table of contents indicates, this topic was interpreted broadly, with discussions of algorithmic techniques, random graphs per se, methods from linear algebra, and other related areas also included.

This monograph describes a method of proof and does so by example. Often giving the "best possible" result conflicted with a clear presentation of the methodology. I have consistently emphasized the methodology; for the best results, the reader may check the original references or examine the problem in more detail first-hand. The following is a new, more encyclopedic work that gives further detail on this method: *The Probabilistic Method,* by Noga Alon and Joel Spencer with an appendix by Paul Erdös, John Wiley, New York, 1992.

To be proficient in the probabilistic method one must have a feel for asymptotic calculations, i.e., which terms are important and which may be safely discarded. The only way I know of to achieve this is to work out many examples in some detail and observe how some seemingly large factors evaporate to $o(1)$ by the final result.

In this second edition, I have tried not to tamper with the lectures. However, the last several years have seen some new developments that I have added. A breakthrough on algorithmic implementation of the Lovász Local Lemma has been made by Jozsef Beck; this has been added to Lecture 8. New results on Zero-One Laws by Saharon Shelah and this author are mentioned in Lecture 3. Additional material on the Double Jump also has been added to Lecture 3. An application to efficient parallel algorithms has been added to Lecture 6.

In the late 1980s Svante Janson discovered probabilistic bounds, now known as the Janson Inequalities, that gave a fundamentally new approach to many problems in Random Graphs. The Bonus Lecture is devoted to these results. It concludes with a proof, using these Inequalities, of Bollobás's celebrated result on the chromatic number of the random graph.

The probabilistic method is not difficult and gives quick strong results to problems that would otherwise be unassailable. I strongly feel that every worker in discrete mathematics should have a facility with at least the basic methodology. The new material in this edition has been added, as much as possible, in separate sections with the original writing left intact. I hope the informal style has been retained and that these lectures remain user-friendly for graduate student and researcher alike.

Finally, the not-so-junior author wishes to thank his wife, Maryann, for her assistance, attendance, encouragement, and understanding. Without her, this enterprise would have little meaning.

LECTURE 1

The Probabilistic Method

Ramsey $R(k, k)$. The probabilistic method is best described by examples. Let us plunge right in. Let $R(k, t)$ denote the Ramsey function, i.e., the minimal n so that if the edges of K_n are two-colored Red and Blue then either there is a Red K_k or a Blue K_t. That $R(k, t)$ is well defined, i.e., its holding, for n sufficiently large, is Ramsey's Theorem, which will not be our concern here. Rather, we shall examine *lower bounds* to the Ramsey function. By simple logic:

> $R(k, t) > n$ means there exists a two-coloring of K_n with neither Red K_k nor Blue K_t.

To show $R(k, t) > n$ we must prove the existence of a coloring of K_n. The diagonal case $R(k, k)$ was first treated by Paul Erdös in 1947. I feel this result inaugurated the probabilistic method. While there are somewhat earlier results, the elegance and influence of this result are unmatched.

THEOREM. *If*

$$\binom{n}{k} 2^{1-\binom{k}{2}} < 1,$$

then $R(k, k) > n$.

Proof. Color K_n *randomly.* More precisely, create a probability space whose elements are Red-Blue colorings of K_n. Define the probabilities by setting

$$\Pr[\chi(i, j) = \text{Red}] = \Pr[\chi(i, j) = \text{Blue}] = \tfrac{1}{2}$$

and letting these events be mutually independent over all edges $\{i, j\}$. Informally, we have a *gedanke experiment* in which a fair coin is flipped to determine the color of each edge.

Let S be a set of k vertices. Let A_S be the *event* that S is monochromatic. Then

$$\Pr[A_S] = 2^{1-\binom{k}{2}}$$

as for S to hold all $\binom{k}{2}$ "coin flips" must be the same. Consider the event $\bigvee A_S$, the disjunction over all $S \in [n]^k$. An exact formula for $\Pr[A_S]$ would be most difficult as the events A_S may have a complex interaction. We use the simple fact that the probability of a disjunction is at most the sum of the probabilities of the events. Thus

$$\Pr[\bigvee A_S] \le \sum \Pr[A_S] = \binom{n}{k} 2^{1-\binom{k}{2}}$$

as there are $\binom{n}{k}$ summands. By assumption this is less than one. Thus $B = \bigwedge \bar{A}_S$ has positive probability. Therefore B is not the null event. Thus there is a point in the probability space for which B holds. But a point in the probability space is precisely a coloring χ of K_n. And the event B is precisely that under this coloring χ there is no monochromatic K_k. Hence $R(k, k) > n$. □

We have proven the existence of a coloring, but we have given no way to find it. Where is the coloring? This state of affairs (or bewilderment, for the neophyte) exemplifies the probabilistic method. For the mathematician reared in the Hilbert formalist school there are no problems here. After all, the probability space is finite, and so the existence of the desired coloring provides no logical difficulties. To others a construction, a polynomial (in n) time algorithm giving a coloring, would be desirable. In future lectures we will occasionally be able to replace a probabilistic proof with an algorithm. For the Ramsey function $R(k, k)$ no construction is known that gives nearly the lower bound that can be derived from the Erdös proof.

In his original paper Erdös used a counting argument for this result, avoiding probabilistic language. Basically, he set Ω equal to the family of all $2^{\binom{n}{2}}$ colorings χ of K_n, and for each k-set S he set A_S equal to the family of colorings for which S is monochromatic, so that

$$|A_S| = 2^{\binom{n}{2} - \binom{k}{2} + 1}.$$

Now since the cardinality of a union is at most the sum of the cardinalities, $|\vee A_S| < |\Omega|$ and so there is a $\chi \in \Omega$ not belonging to any A_S, as desired. Erdös relates that, when presented with the probabilistic method, the probabilist J. Doob remarked, "Well, this is very nice but it's really just a counting argument." With the advantage of hindsight I respectfully disagree. As we shall see (the best example being, perhaps, the Lovász Local Lemma of Lecture 8), notions of probability permeate the probabilistic method. To reduce the arguments to counting, although it is technically feasible, would be to remove the heart of the argument.

Asymptotics. The probabilistic method is generally used when bounds on the asymptotic value of a function are desired. One needs a feeling for these asymptotics.

What is the maximal $n = n(k)$ for which

$$\binom{n}{k} 2^{1-\binom{k}{2}} < 1?$$

Very roughly

$$\binom{n}{k} \sim n^k, \qquad 2^{1-\binom{k}{2}} \sim 2^{-k^2/2}$$

so we want $n^k 2^{-k^2/2} \sim 1$ and $n \sim 2^{k/2}$. Surprisingly, such a very rough estimate is generally sufficient for most purposes. More precisely we approximate

$$\binom{n}{k} = (n)_k/k! \sim n^k/k^k e^{-k} \sim (ne/k)^k.$$

We want

$$(ne/k)^k \sim 2^{k(k-1)/2}$$

so

$$(*) \qquad R(k,k) \geq n \sim (k/e)2^{(k-1)/2} = \frac{k}{e\sqrt{2}} 2^{k/2}.$$

We have thrown away the $\sqrt{2\pi k}$ of Stirling's Formula and 2^{-1} on the right. But these factors disappear when the kth root is taken. Careful analysis shows that (*) is asymptotically correct to within a $(1+o(1))$ factor. (To get a good feeling for the probabilistic method the reader is advised to work out some examples like this in full detail.)

Let

$$f(k) = \binom{n}{k} 2^{1-\binom{k}{2}}$$

and let n_0 be such that $f(n_0) \sim 1$.

One may observe a strong threshold behavior. If $n < n_0(1-\varepsilon)$, then $f(n) \ll 1$. For such n, $\Pr[\bigvee A_S] \leq f(n) \ll 1$ and so a random coloring almost certainly has no monochromatic k_k. Thus for practical purposes there is a simple algorithm for finding a coloring χ. Take $n = n_0(.99)$ and start flipping coins!

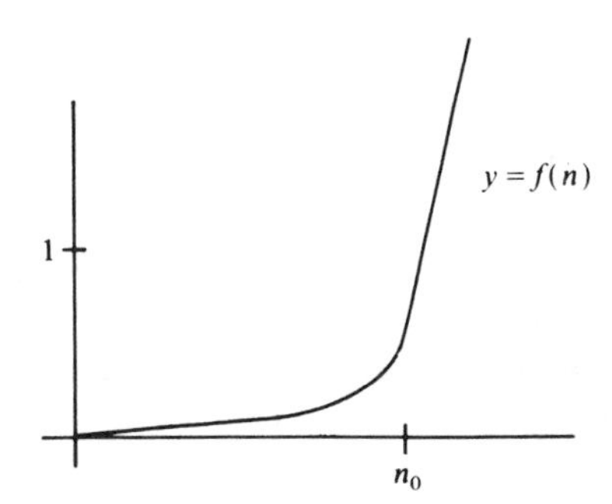

The upper bound for $R(k,k)$ is given by the proof of Ramsey's Theorem.

$$R(k,k) \leq \binom{2k-2}{k-1} \sim c4^k/\sqrt{k}.$$

The value $\lim R(k,k)^{1/k}$ thus lies between $\sqrt{2}$ and 4. Determination of this limit is a major problem in Ramsey Theory and the probabilistic method.

What of the precise values $R(k,k)$? It is known that $R(3,3)=6$, $R(4,4)=17$ and $42 \leq R(5,5) \leq 55$. Is there hope of finding more values? Two anecdotes may suffice.

The values $R(3,3)$ and $R(4,4)$ were found by Greenwood and Gleason in 1955. As Gleason was my advisor, I once spent a weekend puzzling over $R(5,5)$ and then asked him for advice. He was quite clear: "Don't work on it!" Behind their elegant paper lay untold hours of calculation in attempts at improvement. The Law of Small Numbers was at work—simple patterns for k small disappear when k gets too large for easy calculation. Indeed, the value $R(4,5)=25$ was only found in 1993. (There has been more success with $R(3,k)$ with $3 \le k \le 9$ now known.)

Erdös asks us to imagine an alien force, vastly more powerful than us, landing on Earth and demanding the value of $R(5,5)$ or they will destroy our planet. In that case, he claims, we should marshall all our computers and all our mathematicians and attempt to find the value. But suppose, instead, that they ask for $R(6,6)$. In that case, he believes, we should attempt to destroy the aliens.

$R(k,t)$, k fixed. The probabilistic method is easily modified when $k \neq t$.

THEOREM. *If*

$$\binom{n}{k} p^{\binom{k}{2}} + \binom{n}{t} (1-p)^{\binom{t}{2}} < 1$$

for some $p \in [0,1]$, *then* $R(k,t) > n$.

Proof. We color K_n randomly with probabilities given by

$$\Pr[\chi(i,j) = \text{Red}] = p.$$

(In our gedanke experiment the coin is heads with probability p.) For each k-set S let A_S be the event that S is Red, and for each t-set T let B_T be the event that T is Blue. Then

$$\Pr[A_S] = p^{\binom{k}{2}}, \qquad \Pr[B_T] = (1-p)^{\binom{t}{2}}$$

so

$$\Pr[\bigvee A_S \vee \bigvee B_T] \le \sum \Pr[A_S] + \sum \Pr[B_T] = \binom{n}{k} p^{\binom{k}{2}} + \binom{n}{t} (1-p)^{\binom{t}{2}} < 1.$$

With positive probability, χ has neither Red K_k nor Blue K_t. Thus such a coloring exists, completing the proof. □

Let us examine the asymptotics of $R(4,t)$. We want $\binom{n}{4} p^6 = cn^4 p^6 < 1$ so we take $p = \varepsilon n^{-2/3}$. Now we estimate $\binom{n}{t}$ by $n^t(!)$, $1-p$ by e^{-p} and $\binom{t}{2}$ by $t^2/2$, so we want $n^t e^{-pt^2/2} < 1$. Taking tth roots and logs, $pt/2 > \ln n$ and $t > (2/p) \ln n = Kn^{2/3} \ln n$. Expressing n in terms of t

$$R(4,t) > kt^{3/2}/\ln^{3/2} t = t^{3/2+o(1)}.$$

Again the reader may check that more precise calculations, such as not ignoring $t!$ (!), do not greatly affect the end result.

The upper bound given by Ramsey's Theorem is

$$R(k,t) < \binom{t+k-2}{k-1} \sim ct^{k-1}.$$

Actually $o(t^{k-1})$ is known, but the improvement does not change the exponent.

Conjecture. For any fixed k, $R(k, t) = t^{k-1+o(1)}$.

At the Durango Lectures a monetary reward was offered for the resolution of this conjecture.

Tournaments with S_k. The probabilistic method is most striking when the statement of the result does not appear to call for it. A tournament, for the purposes of this monograph, is a complete directed graph on n vertices. That is, there are n players, every pair plays a game and there are no draws. We direct an edge from i to j if i beats j. Note that the schedule of the tournament does not matter, only the results. A tournament has property S_k if for every k players $x_1, \cdots, x_k$ there is some other player y who beats all of them. For example, the

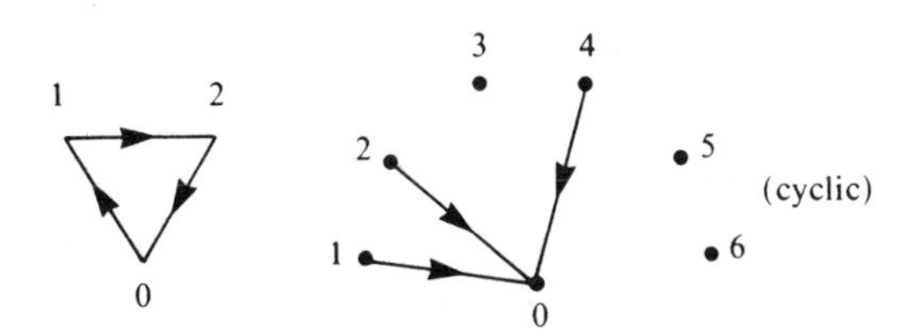

T_3 where 0 beats 1 beats 2 beats 0 has S_1. The T_7 with players Z_7, where i beats j if $i-j$ is a square, has S_2.

THEOREM. *For every k there is a finite T_n with property S_k.*

Proof. Consider a random T on n players, i.e., every game is determined by the flip of a fair coin. For a set X of k players let A_X be the event that no $y \notin X$ beats all of X. Each $y \notin X$ has probability 2^{-k} of beating all of X and there are $n-k$ such y, all of whose chance are mutually independent, so

$$\Pr[A_X] = (1-2^{-k})^{n-k}.$$

Hence

$$\Pr[\bigvee A_X] \leq \binom{n}{k}(1-2^{-k})^{n-k}.$$

Choose n so that

$$(*) \qquad \binom{n}{k}(1-2^{-k})^{n-k} < 1.$$

For this n, $\bigwedge \bar{A}_X$ has positive probability so there is a point in the probability space, i.e., a tournament T, with property S_k.

What are the asymptotics of $(*)$? Roughly we want

$$n^k e^{-2^{-k}n} < 1$$

so we need

$$n > 2^k k^2 (\ln 2)(1+o(1)).$$

In fact, this is the asymptotic bound on $(*)$. Letting $f(k)$ be the smallest number of players in a tournament T with S_k, we find that it is not difficult to show $f(k) > 2^{k-1}$, and the best-known result is $f(k) > ck2^k$.

Extreme tails and ranking tournaments. Let S_n denote the distribution

$$S_n = X_1 + \cdots + X_n$$

where $\Pr[X_i = +1] = \Pr[X_i = -1] = \frac{1}{2}$ and the X_i are mutually independent. Thus $S_n = 2U_n - n$ where U_n is the binomial distribution $B(n, \frac{1}{2})$.

THEOREM. $\Pr[S_n > \lambda\sqrt{n}] < e^{-\lambda^2/2}$ *for all* n, $\lambda \geq 0$.

We prove this result in Lecture 4; however, let us assume it for now. Note that for λ fixed the Central Limit Theorem gives that S_n is approximately normal with mean zero and standard deviation $\sqrt{n}$ so that

$$\lim \Pr[S_n > \lambda\sqrt{n}] = \int_\lambda^\infty \frac{1}{\sqrt{2\pi}} e^{-t^2/2}\, dt$$

which can be shown to be less than $e^{-\lambda^2/2}$.

Given a tournament T on players $1, \cdots, n$ and a ranking σ (i.e., a permutation on $[n]$, $\sigma(i)$ being the rank of player i), let

$$\text{fit}(T, \sigma) = \#[\text{nonupsets}] - \#[\text{upsets}]$$

where a game between i and j is called a nonupset if i beats j and $\sigma(i) < \sigma(j)$; otherwise an upset results. Set

$$\text{fit}(T) = \max \text{fit}(T, \sigma)$$

the maximum over all rankings σ. Note $\text{fit}(T) = \binom{n}{2}$ precisely when T is transitive. Also, $\text{fit}(T) \geq 0$ for any T. Let σ be any ranking and τ the reverse ranking given by $\tau(i) = n + 1 - \sigma(i)$. Upsets in T, σ are nonupsets in T, τ and vice versa, so $\text{fit}(T, \sigma) + \text{fit}(T, \tau) = 0$; therefore one of these terms is nonnegative. Finally, set

$$F(n) = \min \text{fit}(T)$$

over all tournaments T on n players.

THEOREM. $F(n) < n^{3/2}(\ln n)^{1/2}$.

Proof. Let T be a random tournament. For any σ, $\text{fit}(T, \sigma)$ has distribution S_m, where $m = \binom{n}{2}$; after all, each game has independent probability $\frac{1}{2}$ of fitting σ. Let A_σ be the event that $\text{fit}(T, \sigma) > \alpha$. Then

$$\Pr[A_\sigma] < e^{-\alpha^2/2m} < e^{-\alpha^2/n^2} = n^{-n} < 1/n!$$

where we choose $\alpha = n^{3/2}(\ln n)^{1/2}$. This α is $n^{1/2}(\ln n)^{1/2}$ standard deviations off the mean, a place no statistician would go. But in the probabilistic method it is quite common to consider such extreme tails of distributions. Now

$$\Pr[\bigvee A_\sigma] < \sum \Pr[A_\sigma] = n! \Pr[A_\sigma] < 1.$$

We required each individual probability to be less than $1/n!$ to compensate the enormous number $n!$ of possible bad events. With this α, $\bigwedge \bar{A}_\sigma \neq \emptyset$, there is a T with $\bigwedge \bar{A}_\sigma$ so that $\text{fit}(T) < \alpha$. □

Noting that $n^{3/2}(\ln n)^{1/2} = o(n^2)$, let us make a far weaker remark: There is a tournament T such that no ranking σ agrees with 51% of the games. Very often a graph theorist unfamiliar with probabilistic methods and victimized by the Law of Small Numbers will make a conjecture such as "Every tournament has a ranking with 60% of its games correctly ranked." A moderate acquaintance with the probabilistic method is sufficient to send such conjectures to the oblivion they deserve.

Edge discrepancy. Let $g(n)$ be the minimal integer so that if K_n is Red-Blue colored there exists a set of vertices S on which the number of Red edges differs from the number of Blue edges by at least $g(n)$. At first this appears related to the Ramsey function as a monochromatic S will have high discrepancy. It turns out, however, that it is quite large sets S that have high discrepancy even though their ratio of Red to Blue edges is nearly one. In this Lecture we examine only the upper bound on $g(n)$.

THEOREM. $g(n) \le cn^{3/2}$.

Proof. For each set S let A_S be the event that $|\#(\text{Red}) - \#(\text{Blue})| \ge \alpha$ on S. When $|S| = k$, $\#(\text{Red}) - \#(\text{Blue})$ has distribution S_m, where $m = \binom{k}{2} < n^2/2$. Thus

$$\Pr[A_S] < e^{-\alpha^2/n^2}.$$

There are only 2^n possible S so

$$\Pr[\bigvee A_S] < 2^n e^{-\alpha^2/n^2} = 1$$

by taking $\alpha = n^{3/2}(\ln 2)^{1/2}$. $\square$

Exercise. Improve $c = (\ln 2)^{1/2}$ by bounding the number of S of size $n(1-\varepsilon)$. What is the best c you can get?

Linearity of expectation. The property

$$E[\sum X_i] = \sum E[X_i]$$

appears simple, but it is quite useful. Its strength comes from the fact that the X_i need not be independent.

Example. The hat check girl problem. In this old chestnut, thirty men give their hats to the Hat Check Girl and she hands them back at random. On the average, how many men get their own hats back? Let A_i be the event that the ith man gets his hat back and X_i the corresponding indicator random variable. Then $E[X_i] = \Pr[A_i] = 1/n$. Let $X = \sum X_i$, the number of men receiving their own hats back. Then

$$E(X) = \sum E(X_i) = n(1/n) = 1.$$

The actual distribution of X (here roughly Poisson) remains unknown when linearity of expectation is employed.

Example. A Hamiltonian path in a tournament T on players $[n]$ is a permutation σ such that $\sigma(i)$ beats $\sigma(i+1)$, $1 \le i \le n-1$. How many Hamiltonian paths can a tournament have? Let T be random, for each σ let A_σ be the event that σ gives a Hamiltonian path in T and let X_σ be the corresponding indicator random variable. Then

$$E(X_\sigma) = \Pr(A_\sigma) = 2^{-(n-1)}$$

as the $n-1$ random games must be "correct." Set $X = \sum X_\sigma$, the number of Hamiltonian paths. Then

$$E(X) = \sum E(X_\sigma) = n!2^{-(n-1)}.$$

Hence there is a point in the probability space, a specific T, for which X exceeds or equals its expectation. This T has at least $n!2^{-(n-1)}$ Hamiltonian paths.

Addendum. What is the average point value of a bridge hand? We assume Ace $=4$, King $=3$, Queen $=2$, Jack $=1$, Void $=3$, Singleton $=2$, Doubleton $=1$. Let N, S, E, W be the number of high card points in the North, South, etc. hands. Let $T = N+S+E+W$. By symmetry, $E(N) = \cdots = E(W)$ so $E(T) = E(N) + \cdots + E(W) = 4E(N)$. But $T = 40$ always (there are 40 high card points in the deck) so $E(N) = 10$. Now for distribution. Let C, D, P, H be the number of distribution points from Clubs, Diamonds, Spades and Hearts in the North hand. A calculation gives

$$E(C) = 3 \times \Pr(\text{No Clubs}) + 2 \times \Pr(1 \text{ Club}) + 1 \times \Pr(2 \text{ Clubs})$$
$$= 3(.01279) + 2(.08006) + 1(.20587) = .40437.$$

Let $I = C + D + P + H$ be the number of distributional points. Then

$$E(I) = E(C) + \cdots + E(H) = 4E(C) = 1.6175.$$

Finally, let X be the total number of points in the North hand. $X = N + I$, so

$$E(X) = E(N) + E(I) = 11.6175.$$

The objection that "you can't have voids in all four suits" is immaterial; linearity of expectation does not require independence.

Property B. A family F of subsets of Ω has Property B if some Red-Blue coloring of Ω gives no monochromatic $S \in F$. Let $m(n)$ be the least m so that there is an F consisting of m n-sets without Property B. We are interested in lower bounds to $m(n)$. Note that $m(n) > m$ means that any m n-sets can be two-colored so that no set is monochromatic.

THEOREM. $m(n) > 2^{n-1} - 1$.

Proof. Let F have m n-sets and color Ω randomly. For each $S \in F$ let A_S be the event that S is monochromatic. Clearly $\Pr(A_S) = 2^{1-n}$. Then, with $m < 2^{n-1}$,

$$\Pr(\bigvee A_S) \le m2^{1-n} < 1$$

and the desired two-coloring exists. □

van der Waerden. Let $W(k)$ be the least n so that, if $[n]$ is two-colored, there exists a monochromatic arithmetic progression with k terms. The existence of $W(k)$ is the celebrated theorem of van der Waerden. The best upper bound to $W(k)$, due to Saharon Shelah, grows extremely rapidly. We consider only the lower bound.

$W(k) > n$ means $[n]$ may be two-colored so that no k-term arithmetic progressi[illegible] is monochromatic.

THEOREM. $W(k) > 2^{k/2}$.

Proof. Color $[n]$ randomly. For each arithmetic progression with k terms S l[illegible] A_S be the event that S is monochromatic. Then $\Pr(A_S) = 2^{1-k}$. There are le[illegible] than $n^2/2$ such S (as S is determined by its first and second elements—of cours[illegible] this may be slightly improved) so if $(n^2/2)2^{1-k} < 1$ we have

$$\Pr(\bigvee A_S) \le \Pr(A_S) < 1$$

and the desired coloring exists. □

REFERENCES

Three general references to the entire topic of probabilistic methods are:

P. ERDÖS AND J. SPENCER, *Probabilistic Methods in Combinatorics*, Academic Press/Akademiai Kiado, New York-Budapest, 1974.

J. SPENCER, *Nonconstructive methods in discrete mathematics*, in Studies in Combinatorics, G.-C. Rota, ed., Mathematical Association of America, 1978.

———, *Probabilistic methods*, Graphs Combin., 1 (1985), pp. 357-382.

Another general article on this topic appears in the *Handbook of Combinatorics* to be published in 1994 by North-Holland.

The three-page paper that began this subject by giving $R(k, k)$ is:

P. ERDÖS, *Some remarks on the theory of graphs*, Bull. Amer. Math. Soc., 53 (1947), pp. 292-294.

Tournaments with S_k:

P. ERDÖS, *On a problem of graph theory*, Math. Gaz., 47 (1963), pp. 220-223.

Tournament ranking:

P. ERDÖS AND J. MOON, *On sets of consistent arcs in a tournament*, Canad. Math. Bull., 8 (1965), pp. 269-271.

Property B:

P. ERDÖS, *On a combinatorial problem*, I, Nordisk Tidskr. Infomations-behandlung (BIT), 11 (1963), pp. 5-10.

The combinatorial papers of Paul Erdös, including the above four, have been collected in:

Paul Erdös: The Art of Counting, J. Spencer, ed., MIT Press, Cambridge, MA, 1973.

LECTURE 2

The Deletion Method and Other Refinements

Ramsey $R(k, k)$. Often a desired configuration can be found by taking a random configuration and making a "small" modification.

THEOREM.

$$R(k,k) > n - \binom{n}{k} 2^{1-\binom{k}{2}}.$$

Proof. Randomly color K_n. Let X be the number of monochromatic k-sets. Then $X = \sum X_S$, the sum over all k-sets S, where X_S is the indicator random variable of the event A_S that S is monochromatic.

$$E(X_S) = \Pr(A_S) = 2^{1-\binom{k}{2}}.$$

By linearity of expectation

$$E(X) = \sum E(X_S) = \binom{n}{k} 2^{1-\binom{k}{2}}.$$

There is a point in the probability space for which X does not exceed its expectation. That is, there is a coloring with at most

$$\binom{n}{k} 2^{1-\binom{k}{2}}$$

monochromatic S. Fix that coloring. For each monochromatic S select a point $x \in S$ arbitrarily and *delete* it from the vertex set. The remaining points V^* have no monochromatic k-set and

$$|V^*| \geq n - \binom{n}{k} 2^{1-\binom{k}{2}}. \qquad \square$$

What about the asymptotics? We should choose n so that

$$\binom{n}{k} 2^{1-\binom{k}{2}} \ll n.$$

That is,

$$n^{k-1} \ll k! 2^{k(k-1)/2}.$$

Taking $(k-1)$st roots

$$n<(k/e)2^{k/2}(1+o(1)).$$

For such n the deleted points will be negligible and so

$$R(k,k)\gtrsim n\sim(k/e)2^{k/2}.$$

This is an improvement over the previous bound by a $\sqrt{2}$ factor. Of course, this is negligible when viewed in light of the gap between the upper and lower bounds, but we do what we can.

Off-diagonal Ramsey numbers. By the same argument,

$$R(k,t)>n-\binom{n}{k}p^{\binom{k}{2}}-\binom{n}{t}(1-p)^{\binom{t}{2}}$$

for any n, p, $0\le p\le 1$. Let us examine the asymptotics when $k=4$. We want $n^4p^6\ll n$ so we select $p=\varepsilon n^{-1/2}$. Then

$$\binom{n}{t}(1-p)^{\binom{t}{2}}<n^t e^{-pt^2/2}\ll 1$$

when

$$t>\frac{2}{p}\ln n=Kn^{1/2}\ln n.$$

If we express n in terms of t,

$$R(4,t)>ct^2/\ln^2 t=t^{2+o(1)}.$$

Exercise. Show $R(3,t)>t^{3/2+o(1)}$ using the deletion method. Note how the usual probabilistic method is ineffective in bounding $R(3,t)$.

Turan's theorem. Let us express the celebrated result of Paul Turan in a form convenient for us. If G has n vertices and $nk/2$ edges, then $\alpha(G)\ge n/(k+1)$. We can come "halfway" to Turan's Theorem with the deletion method.

THEOREM. *If G has n vertices and $nk/2$ edges then $\alpha(G)\ge n/2k$.*

Proof. Let V, E denote the vertex and edge sets, respectively, of G. Select a subset $S\subseteq V$ randomly by letting

$$\Pr[x\in S]=p$$

with these probabilities mutually independent over all $x\in V$. In our gedanke experiment we take a coin which comes up heads with probability p and flip it for each $x\in V$ to see if x is "chosen" to be in S. Let X denote the number of vertices of S and let Y denote the number of edges of G in S. Clearly (exercise):

$$E(X)=np.$$

For each $e\in E$ let Y_e be the indicator random variable for the event $e\subseteq S$. Then

$E[Y_e]=p^2$ as for e to be inside S two vertices must be chosen. As

$$Y=\sum_{e\in E} Y_e, \qquad E(Y)=\sum_{e\in E} E(Y_e)=\binom{nk}{2}p^2,$$

then

$$E(X-Y)=E(X)-E(Y)=np-(nk/2)p^2.$$

Let us choose $p=1/k$ to maximize the above expression:

$$E(X-Y)=n/k.$$

There is a point in the probability set for which $X-Y$ is at least n/k. That is, there is a set S which has at least n/k more vertices than edges. Delete one point from each edge from S leaving a set S^*. Then S^* is independent and has at least n/k points.

In a formal proof the variable p would not appear; S would be defined by $\Pr[x\in S]=1/k$. The neophyte, reading such a definition in the opening line of a proof in a research paper, might be somewhat mystified as to where the "$1/k$" term came from. Rest assured, this value was a variable when the result was being found. One way to understand a probabilistic proof is to reset all the probabilities to variables and see that the author has, indeed, selected the optimal values.

Property B. Let us return to the Property B function $m(n)$ of Lecture 1. Suppose $m=2^{n-1}k$ and F is a family of m n-sets with Ω the underlying point set. We give a recoloring argument of J. Beck that shows Ω may be two-colored without monochromatic $S\in F$ as long as $k<n^{1/3-o(1)}$.

First we color Ω randomly. This gives X monochromatic $S\in F$ where $E(X)=k$. Now we *recolor*. For every $x\in\Omega$ which lies in some monochromatic SF we change the color of x with probability $p=[(\ln k)/n](1+\varepsilon)$. We call these the First and the Second Coloring or Stage. To be precise, even if x lies in many monochromatic $S\in F$ the probability of its color changing is just p.

Given that S was monochromatic at the First Stage the probability that it remained so was $(1-p)^n+p^n\sim(1-p)^n\sim e^{-pn}=k^{-1-\varepsilon}$. The expected number of S monochromatic at both stages is then $k(k^{-1-\varepsilon})=k^{-\varepsilon}$, which is negligible. Of course, p was selected just large enough to destroy the monochromatic S. Has the cure been too strong, so that new monochromatic T have appeared?

For each $S, T\in F$ with $S\cap T\neq\varnothing$ let A_{ST} be the event that S was Red at Stage 1 and T became Blue at Stage 2. That is, in curing S we have created T. We will bound $\Pr[A_{ST}]$ in the case $|S\cap T|=1$, say $S\cap T=\{x\}$. It can be shown (exercise) that $\Pr[A_{ST}]$ is even smaller when $|S\cap T|$ is larger. A_{ST} can occur if, for instance, the following conditions hold:

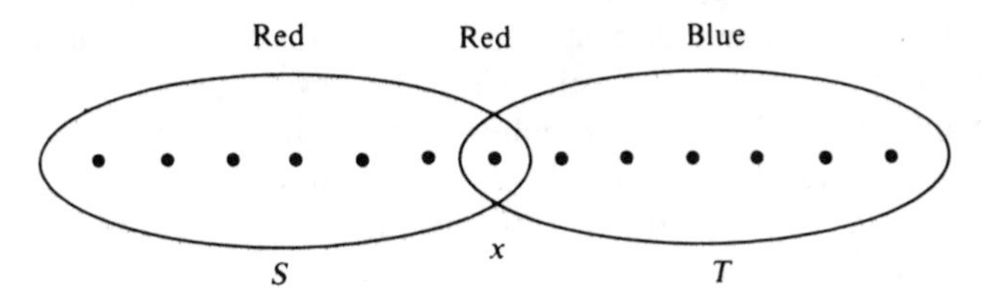

This occurs with probability $2^{-(2n-1)}p$ since the First Coloring of $S \cup T$ is determined and then x must change color. More generally, though, for $V \subseteq T - S$ with $|V| = v$, let A_{STV} be the following event:

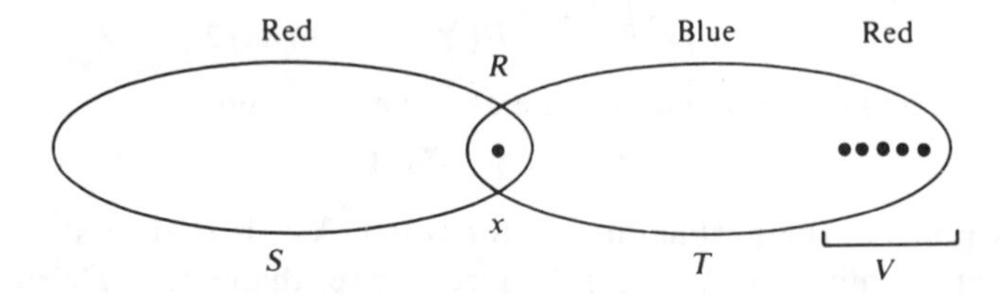

$S \cup V$ is Red at Stage 1, the rest of T is Blue at Stage 2, and $V \cup \{x\}$ changes color at Stage 2. As the color of $S \cup T$ is determined,

$$\Pr[A_{STV}] = 2^{-(2n-1)} \Pr[V \cup \{x\} \text{ becomes Blue} \mid \text{First Coloring of } S \cup T].$$

This conditional probability is at most p^{v+1} as each of the $v+1$ points $V \cup \{x\}$ must turn Blue. (Perhaps it is much less than that, as each $y \in V$ must also lie in a Red set. I feel that, if we could take full advantage of this additional criterion, the bound on $m(n)$ could be improved.) Thus

$$\Pr[A_{STV}] \leq 2^{-(2n-1)} p^{v}.$$

Hence

$$\begin{aligned} \Pr[A_{ST}] &\leq \sum_{V} \Pr[A_{STV}] \leq \sum_{v=0}^{n-1} \binom{n-1}{v} 2^{-(2n-1)} p^{v+1} \\ &= p2^{1-2n}(1+p)^{n-1} \qquad \text{(Binomial Theorem)} \\ &\sim cp2^{-2n} e^{pn} \sim cp2^{-2n} k^{1+\varepsilon}. \end{aligned}$$

(We can calculate that the main contribution to the sum occurs not at $v = 0$ but rather near $v = \ln k$. Can we use that these $\ln k$ points all lie in Red sets at Stage 1?) There are at most $(2^{n-1}k)^2 = c2^{2n}k^2$ choices of S, T so

$$\begin{aligned} \Pr[\vee A_{ST}] &\leq c(2^{2n}k^2)(2^{-2n}k^{1+\varepsilon}p) \\ &= ck^{3+\varepsilon}p = ck^{3+\varepsilon}(\ln k)/n \ll 1 \end{aligned}$$

if $k = n^{1/3-o(1)}$. That is, for this k the expected number of monochromatic sets at Stage 2 is much less than one so with positive probability there are no monochromatic sets. Thus there exists a coloring of Ω with no monochromatic sets. Thus $m(n) > c2^n n^{1/3-o(1)}$. In fact, with a bit more care, $m(n) > c2^n n^{1/3}$ may be shown.

Erdös has shown the upper bound $m(n) < c2^n n^2$. It may appear then that the bounds on $m(n)$ are close together. But from a probabilistic standpoint a factor of 2^{n-1} may be considered as a unit. We could rewrite the problem as follows. Given a family F let X denote the number of monochromatic sets under a random coloring. What is the maximal $k = k(n)$ so that, if F is a family of n-sets with $E(X) \leq k$, then $\Pr[X = 0] > 0$? In this formulation $cn^{1/3} < k(n) < cn^2$ and the problem clearly deserves more attention.

Tournament ranking. Let us return to the tournament ranking function $F(n)$ of Lecture 1.

THEOREM. $F(n) \le cn^{3/2}$, $c = 3.5$.

While I have been emphasizing methodology, there is plenty of room in combinatorics for just plain cleverness. I think this proof, due to Ferdinand de la Vega, is a real gem! De la Vega actually shows that if T is a random tournament then, almost always, fit $(T, \sigma) \le cn^{3/2}$ for all rankings σ. Please review the proof in Lecture 1 and be guided by the notion that if there were only K^n rankings, then the result would be immediate.

For any disjoint sets A, B of players let

$$G(A, B) = \#[\text{nonupsets}] - \#[\text{upsets}]$$

where a game between $a \in A$ and $b \in B$ is called a nonupset if a beats b. Assume $n = 2^t$, though this is only a convenience. Set

$$A_1 = \{i: 1 \le \sigma(i) \le n/2\}, \qquad A_2 = \{i: n/2 < \sigma(i) \le n\}.$$

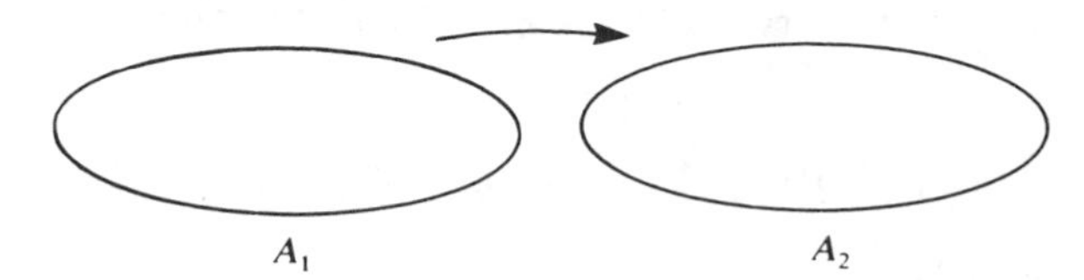

Consider the statement

$$S_1: \quad \text{For every } \sigma,\ G(A_1, A_2) < \sqrt{n^2/4}\sqrt{n}\sqrt{2\ln 2}(1+\varepsilon).$$

Here $\varepsilon > 0$ is arbitrarily small but fixed. I claim that S_1 holds almost always. For any (A_1, A_2), $G(A_1, A_2)$ has distribution S_m where $m = |A_1||A_2| = n^2/4$. Thus

$$\Pr[G(A_1, A_2) > \alpha\sqrt{n^2/4}] < e^{-\alpha^2/2}.$$

The key observation here is that there are only $\binom{n}{n/2} \le 2^n$ possible (A_1, A_2) as opposed to $n!$ possible σ. Thus

$$\Pr[\bar{S}_1] \le \sum_{A_1, A_2} \Pr[G(A_1, A_2) > \alpha\sqrt{n^2/4}] \le 2^n e^{-\alpha^2/2} \ll 1$$

where $\alpha = \sqrt{n}\sqrt{2\ln 2}(1+\varepsilon)$. We have "taken care of" games between the top and bottom halves of the players under any ranking. Inside each half we now examine the quartiles. Set

$$A_j = \{i: (j-1)(n/4) < \sigma(i) \le j(n/4)\}, \qquad 1 \le j \le 4.$$

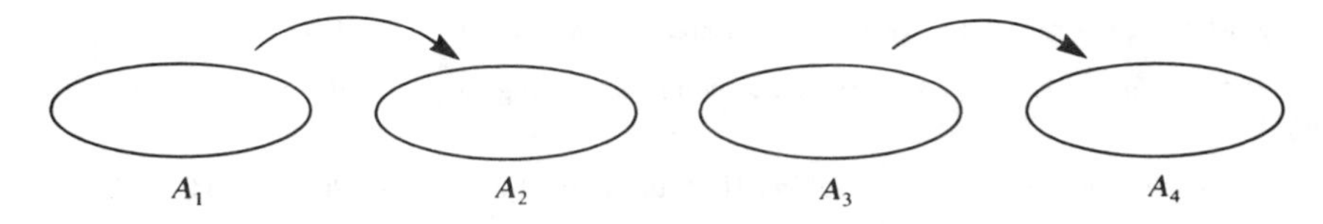

Consider the statement:

$$S_2: \quad \text{For every } G(A_1, A_2) + G(A_3, A_4) < \sqrt{n^2/8}\sqrt{n}\sqrt{2\ln 4}(1+\varepsilon).$$

I claim that S_2 holds almost always. The argument is almost identical. Now there are less than 4^n possible (A_1, A_2, A_3, A_4). For each one $G(A_1, A_2)+G(A_3, A_4)$ has distribution S_m, where $m=|A_1||A_2|+|A_3||A_4|=n^2/8$ and $\Pr[S_m>\alpha\sqrt{m}]< e^{-\alpha^2/2} \ll 4^{-n}$ when $\alpha=\sqrt{n}\sqrt{2\ln 4}(1+\varepsilon)$.

In general, for $1 \le s \le t$ split the players into 2^s groups. That is, given σ let

$$A_j=\{i: (j-1)n2^{-s}<\sigma(i)\le jn2^{-s}\}, \qquad 1\le j\le 2^s$$

and set

$$G^{(s)}=\sum_{j=1}^{2^{s-1}} G(A_{2j-1}, A_{2j}).$$

Then $G^{(s)}$ has distribution S_m, where $m=2^{s-1}(n2^{-s})^2=n^2/2^{s+1}$. There are at most $(2^s)^n$ possible partitions $(A_1, \ldots, A_{2^s})$. (The number of partitions is increasing with s, but this will be more than compensated for by the fact that the number of games being considered is decreasing.) Consider the statement

$$S_s\text{:}\quad \text{For every } \sigma\ G^{(s)}\le\sqrt{n^2/2^{s+1}}\sqrt{n}\sqrt{2\ln 2^s}(1+\varepsilon).$$

Then

$$\Pr[\bar{S}_s]\le 2^{sn}\Pr[S_m\ge\sqrt{m}\sqrt{n}\sqrt{2\ln 2^s}(1+\varepsilon)]\ll 1.$$

With slightly more care,

$$\Pr\left[\bigvee_{s=1}^{t}\bar{S}_s\right]\le\sum_{s=1}^{t}\Pr[\bar{S}_s]\ll 1.$$

That is, a random T satisfies $\bigwedge S_s$. For such T given any σ

$$\begin{aligned}\text{fit}(T,\sigma)&=\sum_{s=1}^{t}G^{(s)}\\&=n^{3/2}(1+\varepsilon)\sum_{s=1}^{t}\sqrt{2\ln 2^s}/\sqrt{2^{s+1}}\\&<3.5n^{3/2}.\end{aligned}$$

REFERENCES

The lower bound on $m(n)$:

J. BECK, *On 3-chromatic hypergraphs*, Discrete Math., 24 (1978), pp. 127–137.

The upper bound on $m(n)$ is an interesting example of standing the probabilistic method on its head. See:

P. ERDÖS, *On a combinatorial problem* II, Acta Math. Hungar., 15 (1964), pp. 445–447.

The removal of $(\ln n)^{1/2}$ in ranking tournaments:

W. F. DE LA VEGA, *On the maximal cardinality of a consistent set of arcs in a random tournament*, J. Combin. Theory, Ser. B, 35 (1983), pp. 328–332.

LECTURE 3

Random Graphs I

Threshold Functions. Here are three models for random graphs. None is the "correct" view; it is best to move comfortably from one to the other. In all cases G has n vertices.

Dynamic. Imagine G to have no edges at time 0; at each time unit a randomly chosen edge is added to G. Then G *evolves* from empty to full.

Static. Given e, let G be chosen randomly from among all graphs with e edges.

Probabilistic. Given p, let the distribution of G be defined by

$$\Pr[\{i, j\} \in G] = p$$

for all i, j with these probabilities mutually independent. That is, flip a coin, heads with probability p, to determine if each edge is in G.

When $p = e/\binom{n}{2}$ the Static and Probabilistic models are nearly identical. We work with the Probabilistic model as it is technically much easier. This model is denoted $G_{n,p}$.

Let A be a property of graphs, assumed monotone for convenience. A function $p(n)$ is called a *threshold function* for A if

(i) $\lim r(n)/p(n) = 0$ implies $\lim \Pr[G_{n,r(n)}$ has $A] = 0$;

(ii) $\lim r(n)/p(n) = 1$ implies $\lim \Pr[G_{n,r(n)}$ has $A] = 1$.

Set $f(r) = f_A(r) = \Pr[G_{n,r}$ has $A]$. When $p(n)$ is a threshold function the function $f(r)$ jumps from zero to one around $r = p$. In the Dynamic model the graph G almost certainly does not have property A when there are $\ll p(n)n^2$ edges and almost certainly does have property A when there are $\gg p(n)n^2$ edges, so G attains property A at some time near $p(n)n^2$.

The theory of random graphs was introduced by Paul Erdös and Alfréd Rényi. They observed that for many natural properties A there is a threshold function $p(n)$. Some examples follow:

PROPERTY:	THRESHOLD:
Contains path of length k	$p(n) = n^{-(k+1)/k}$,
Is not planar	$p(n) = 1/n$,
Contains a Hamiltonian path	$p(n) = (\ln n)/n$,
Is connected	$p(n) = (\ln n)/n$,
Contains a clique on k points	$p(n) = n^{-2/(k-1)}$.

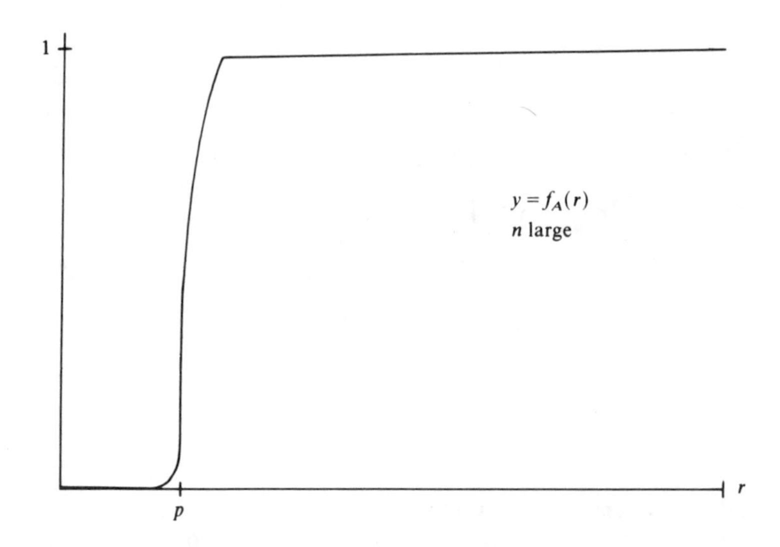

In many cases a much stronger threshold phenomenon was observed. For instance, in the second, third and fourth examples above, if $r(n) < p(n)(1-\varepsilon)$, then the property almost certainly does not hold, while it does hold if $r(n) > p(n)(1+\varepsilon)$, an arbitrarily small constant. Let us examine the final example with $k = 4$ in some detail.

For each 4-set S let A_S be the event that S is a clique and let X_S be the associated indicator random variable so that

$$E[A_S] = \Pr[A_S] = p^6.$$

Let X be the number of K_4's so that

$$X = \sum X_S,$$

the summation over all 4-sets S. By linearity of expectation,

$$E[X] = \sum E[X_S] = \binom{n}{4} p^6 \sim cn^4 p^6.$$

Now $E(X) \sim 1$ when $p \sim kn^{-2/3}$, so it is natural to try to show that (as will be the case) $p(n) = n^{-2/3}$ is a threshold function.

One part is easy. If $p(n) \ll n^{-2/3}$, then $E(X) \ll 1$ and

$$\Pr[X > 0] \leq E[X] \ll 1$$

so that almost certainly G does not contain a K_4. Now assume $p(n) \gg n^{-2/3}$. The same argument gives $E[X] \gg 1$. We would like the following:

$$(??) \qquad E[X] \gg 1 \text{ implies } \Pr[X = 0] \ll 1.$$

But beware! Implication (??) does *not* hold in general. To give this implication (when we can) requires a new technique.

The second moment method. A basic inequality of Chebyshev states that if X has mean m and variance σ^2 and $\lambda \ge 0$ then

$$\Pr[|X-m| \ge \lambda\sigma] \le \lambda^{-2}.$$

With $\lambda = m/\sigma$,

$$\Pr[X=0] \le \Pr[|X-m| \ge m] \le \sigma^2/m^2.$$

For our purposes,

$$(*) \qquad \text{If } E[X] \to \infty \text{ and } \operatorname{var}(X) = o(E(X)^2), \text{ then } \Pr[X=0] \to 0.$$

(The limits reflect the hidden variable n, the number of points.) This use of (*) we call the second moment method. We are only interested in cases where $E[X] \to \infty$; the hard part is to show that $\operatorname{var}(X) = o(E[X]^2)$. As $X = \sum X_S$, we write

$$\operatorname{var}(X) = \sum_{S,T} \operatorname{cov}(X_S, X_T).$$

In this case, and often when the second moment method is employed, X is the sum of m indicator random variables each with expectation μ. Here

$$m = \binom{n}{4} \sim cn^4 \quad \text{and} \quad \mu = p^6.$$

$$\begin{aligned}\operatorname{cov}(X_S, X_T) &= E(X_S X_T) - E(X_S)E(X_T)\\ &= E(X_T \mid X_S = 1)E(X_S) - E(X_S)E(X_T)\\ &= \mu^2 f(S, T)\end{aligned}$$

where we set

$$f(S, T) = \frac{E[X_T \mid X_S = 1]}{E(X_T)} - 1.$$

As there are m^2 choices of S, T

$$\begin{aligned}\operatorname{var}(X) &= m^2\mu^2 E_{S,T}[f(S, T)]\\ &= E[X]^2 E_{S,T}[f(S, T)]\end{aligned}$$

where by the second factor we mean the expected value of $f(S, T)$ when S, T are chosen randomly from among all 4-sets. Then (*) becomes

$$E(X) \to \infty, \qquad E_{S,T}[f(S, T)] = o(1) \Rightarrow \Pr[X=0] \to 0.$$

As all S are symmetric it is convenient to consider S fixed (say $S = \{1, 2, 3, 4\}$) and remove it as a variable from f. We must check whether

$$(?) \qquad E_T[f(T)] = o(1).$$

The value $f(T)$ depends only on $i = |S \cap T|$. Thus

$$E_T[f(T)] = \sum_{i=0}^{4} \Pr[|S \cap T| = i] \times [f(S, T) \text{ when } |S \cap T| = i].$$

We need show that each summand is small.

Case (i): $i=0$ *or* 1. The events A_S, A_T are then independent, as they involve disjoint edges. (Here the convenience of the Probabilistic versus the Static model is apparent.) Hence

$$E[X_T \mid X_S=1]=E[X_T]$$

and $f(T)=0$.

In all other cases it is simpler to look at $f(T)+1$.

Case (ii): $i=4$. That is, $S=T$. $\Pr[S=T]=1/m$ and

$$1+f(S)=\frac{E[X_S \mid X_S=1]}{\mu}=\frac{1}{\mu}$$

so

$$\Pr[T=S]f(S)\leq 1/m\mu=1/E(X)=o(1).$$

This will always be the situation with $S=T$. Quite generally

$$\operatorname{var}(X)=\sum \operatorname{var}(X_S)+\sum_{S\neq T}\operatorname{cov}(X_S, X_T).$$

As X_S is 0 to 1, $\operatorname{var}(X_S)\leq E[X_S]$ and

$$\sum \operatorname{var}[X_S]\leq \sum E[X_S]=E[X]=o(E[X]^2).$$

Case (iii): $i=2$. Here $1+f(T)=p^{-1}$ since conditioning on A_S places one edge

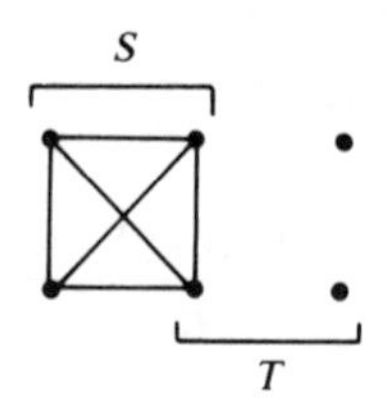

in T and raises $\Pr[A_T]$ from p^6 to p^5. But

$$\Pr[|S\cap T|=2]=\binom{4}{2}\binom{n-4}{2}\Big/\binom{n}{4}\sim cn^{-2}.$$

Thus

$$\Pr[|S\cap T|=2]\times f(T)\sim cn^{-2}p^{-1}\ll 1$$

since $p\sim n^{-2/3}$.

Case (iv): $i=3$. Here $1+f(T)=p^{-3}$ as three edges have been placed in T. But $\Pr[|S\cap T|=3]\sim cn^{-3}$, so the product $cn^{-3}p^{-3}\ll 1$.

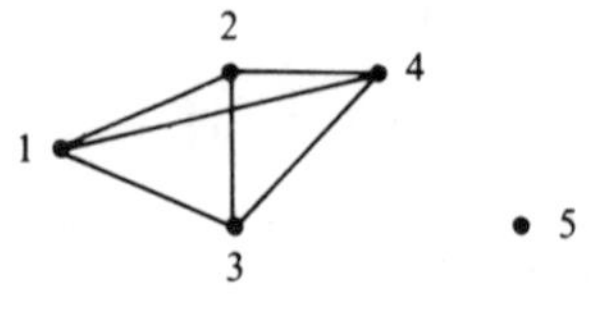

$S=\{1, 2, 3, 4\}$,
$T=\{2, 3, 4, 5\}$

We summarize with Table 1, giving upper bounds on $f(T)$ and the probability of having such T.

TABLE 1

$\|S \cap T\|$	$f(T)$	Pr
0, 1	0	1
2	p^{-1}	n^{-2}
3	p^{-3}	n^{-3}
4	p^{-6}	n^{-4}

All products are $o(1)$, so $E_T[f(T)] = o(1)$ and hence the second moment method applies: $\Pr[X=0] \sim 0$ and so $p = n^{-2/3}$ is indeed the threshold function.

Exercise. Let H be K_4 with a tail and X the number of H's in $G_{n,p}$. As $E[X] = cn^5p^7$, we might think $p = n^{-5/7}$ is the threshold function. Actually this

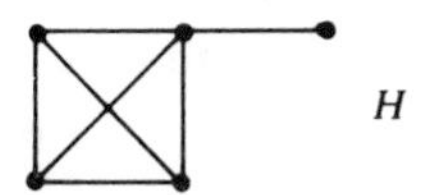

cannot be so since K_4's do not appear until $p = n^{-2/3} \gg n^{-5/7}$. Apply the second moment method to X and see what goes "wrong." More generally, for any H let X_H be the number of H's. We call "balanced" those H's whose threshold function for containing an H is that p for which $E[X_H] = 1$; we call the others "unbalanced." Can we express "H balanced" in graph-theoretic terms?

Poisson approximation. Let us stay with K_4 and examine the situation more closely when $p = cn^{-2/3}$ (c is an arbitrary positive constant). Then

$$E[X] = m\mu \sim c^6/24 = k,$$

a constant. What is the distribution of X? If the X_S were mutually independent then X would have binomial distribution $B(m, \mu)$, which is approximately Poisson k. That will be the case here though, of course, sometimes the X_S are dependent. By inclusion-exclusion

$$\Pr[X=0] = 1 - \sum E[X_S] + \sum E[X_{S_1}X_{S_2}] - \cdots \pm F^{(r)} \mp \cdots$$

where

$$F^{(r)} = \sum E[X_{S_1} \cdots X_{S_r}],$$

the sum over all distinct $S_1, \cdots, S_r$. There are $\binom{m}{r} \sim m^r/r!$ terms. If the X_S were independent each term would be μ^r and then

$$(*) \qquad F^{(r)} \to \binom{m}{r}\mu^r \to \frac{k^r}{r!}.$$

To show $(*)$ we need essentially to show that addends of $F^{(r)}$ for which S_i have dependency have a negligible effect. For $r=2$ this analysis was mainly given in the last section. (One may note $F^{(2)}=\frac{1}{2}(E[X^2]-E[X])$ to get a direct translation.) For general r the argument for (*) is technically more complicated but in the same spirit. We do not give the argument here but show its sufficiency.

THEOREM. *If $F^{(r)}\to k^r/r!$ for all r, then X is asymptotically Poisson in that for all $t\ge 0$, $\Pr[X=t]\to e^{-k}k^t/t!$*

Proof. We only do the case $t=0$. Setting $F^{(0)}=1$, we have

$$\Pr[X=0]=\sum_{r=0}^{\infty}(-1)^r F^{(r)}$$

where $F^{(r)}\to k^r/r!$ for each r. But beware! The limit of an infinite sum is not necessarily the sum of the limits. We apply the Bonferroni Inequalities, which state, in general, that

$$\sum_{r=0}^{2s+1}(-1)^r F^{(r)}\le \Pr[X=0]\le \sum_{r=0}^{2s}(-1)^r F^{(r)}.$$

(That is, the inclusion-exclusion formula alternately under- and overestimates the final answer.) Let $\varepsilon>0$. Choose s so that

$$\left|\sum_{r=0}^{2s}(-1)^r\frac{k^r}{r!}-e^{-r}\right|<\frac{\varepsilon}{2}.$$

Choose n (the hidden variable) so large that for $0\le r\le 2s$

$$|F^{(r)}-k^r/r!|<\varepsilon/2(2s+1).$$

For this n

$$\left|\sum_{r=0}^{2s}(-1)^r F^{(r)}-e^{-r}\right|<\varepsilon$$

and therefore

$$\Pr[X=0]<e^{-r}+\varepsilon.$$

The lower bound on $\Pr[X=0]$ is similar. As ε is arbitrarily small $\Pr[X=0]\to e^{-r}$. □

The zero-one law. Let us restrict ourselves to the First Order Theory of Graphs. The language is simply Boolean connectives, existential and universal quantifiers, variables, equality and adjacency (written $I(x,y)$). The axioms are simply $(x)\sim I(x,x)$ and $(x)(y)I(x,y)\equiv I(y,x)$. Here are some things we can say in this language:

(a) There is a path of length 3

$$(Ex)(Ey)(Ez)(Ew)I(x,y)\wedge I(y,z)\wedge I(z,w).$$

(b) There are no isolated points

$$(x)(Ey)I(x,y).$$

(c) Every triangle is contained in a K_4

$$(x)(y)(z)((I(x, y) \wedge I(x, z) \wedge I(y, z)) \Rightarrow (Ew)(I(x, w) \wedge I(y, w) \wedge I(z, w))).$$

Many other statements cannot be made. For example the following are not expressible: G is connected, G is planar, G is Hamiltonian. Fix p, $0<p<1$. (It does not really matter, so suppose $p=\frac{1}{2}$.)

THEOREM. *For every first order statement A*

$$\lim_{n\to\infty} \Pr[G_{n,p} \text{ has } A]=0 \text{ or } 1.$$

Proof. The proof is a beautiful blend of graph theory, probability and logic. For every r, s let $A_{r,s}$ be the statement "Given any distinct $x_1, \cdots, x_r, y_1, \cdots, y_s$ there is a z adjacent to all of the x_i and none of the y_j." For each r, s (exercise)

$$\lim_{n} \Pr[G_{n,p} \text{ has } A_{r,s}]=1.$$

(Hint: Compare with tournaments with S_k in Lecture 1.)

Let G, G^* be two countable graphs (say, for convenience, each on vertex set N, the positive integers) both satisfying $A_{r,s}$ for all r, s. We claim G and G^* are isomorphic! Let us define an isomorphism $f: G \to G^*$ by determining $f(1), f^{-1}(1)$, $f(2), f^{-1}(2), \cdots$ in order. (To start set $f(1)=1$.) When $f(i)$ needs to be determined, f has been determined on a finite set V. We set $f(i)$ equal to any $y \in G^*$ with the property that if $v \in V$ and $\{v, i\} \in G$ then $\{f(v), y\} \in G^*$, and if $\{v, i\} \notin G$ then $\{f(v), y\} \notin G^*$. Since G^* has all properties $A_{r,s}$ and V is finite there is such a y. We define $f^{-1}(i)$ analogously. Of course, if for some y, $f^{-1}(y)$ had already been defined as i then simply $f(i)=y$. After this countable procedure f is an isomorphism from G to G^*.

Now I claim the system consisting of all $A_{r,s}$ is *complete*, i.e., for any B either B or $\sim B$ is provable from the $A_{r,s}$. Suppose this were false for a particular B. Then adding B gives a theory T and adding $\sim B$ gives a theory T^*, both of which are consistent. But the Gödel Completeness Theorem says that consistent theories must have countable models (or finite models, but the $A_{r,s}$ force a model to be infinite) so they would have countable models G, G^*. We just showed G and G^* must be isomorphic; thus they cannot disagree on B.

Let A now be any first order statement. Suppose A is provable from the $A_{r,s}$. Since proofs are finite, A is provable from a *finite* number of $A_{r,s}$. Then

$$\Pr[\sim A \text{ in } G_{np}] \leq \textstyle\sum^* \Pr[\sim A_{r,s} \text{ in } G_{np}]$$

where $\sum^*$ is over the finite set of $A_{r,s}$ used to prove A. As n approaches infinity each of the addends approaches zero and hence the finite sum approaches zero, $\Pr[\sim A] \to 0$ and $\Pr[A] \to 1$. If A is not provable then (completeness) $\sim A$ is, $\Pr[\sim A] \to 1$ and $\Pr[A] \to 0$. □

For the student of random graphs the situation "p an arbitrary constant" is only one case. What happens when $p=p(n)$, a function tending to zero? For which $p(n)$ can a Zero-One Law be given? Some $p(n)$ do not work. For example, if $p(n)=n^{-2/3}$ and A is "there exists K_4" then the Poisson approximation gives $\lim \Pr[G_{n,p} \text{ has } A]=1-e^{-1/24}$. Roughly it seems that those $p(n)$ for which a

Zero-One Law does not hold are just the threshold functions for first order A. When $p(n) \ll 1/n$ we can characterize those $p(n)$ for which a Zero-One Law holds.

THEOREM. *Let k be an arbitrary positive integer and assume*

$$n^{-1-1/k} \ll p(n) \ll n^{-1-1/(k+1)}.$$

Then for any first order statement A

$$\lim_{n\to\infty} \Pr[G_{n,p} \text{ has } A] = 0 \text{ or } 1.$$

Proof. We imitate the proof techniques used for p constant. Consider the statements:

B: There are no $k+2$ points containing a spanning tree;

C: There is no cycle on $\le k+1$ points;

and the schema

$A_{T,r}$: There exist r components T where T ranges over all trees with at most $k+1$ points (including the "one point tree," an isolated point) and r ranges over all positive integers. For example, when $r=1$ and T is an edge we write

$A_{T,r}$: $(Ex)(Ey)(I(x,y) \wedge (z)((z \ne x \wedge z \ne y) \Rightarrow ((\sim I(x,z)) \wedge (\sim I(y,z)))))$. We outline the arguments that all of these statements hold with probability approaching one.

B: There are $\le n^{k+2}$ sets of $k+2$ elements and c_k spanning trees on a given $k+2$ elements. A given tree is in $G_{n,p}$ with probability p^{k+1}. Thus

$$\Pr[\bar{B}] \le c_k n^{k+2} p^{k+1} \ll 1$$

since $p \ll n^{-1-1/(k+1)}$.

C: For each $t \le k+1$ there are $< n^t$ possible cycles of length t each of which is in $G_{n,p}$ with probability p^t. Thus

$$\Pr[\bar{C}] \le \sum_{t=3}^{k+1} n^t p^t \ll 1.$$

(We needed only $p \ll 1/n$ here.)

C: Fix a tree T on $s \le k+1$ points. For each set S of s vertices, let A_S be the event that $G|_S \cong T$ with this T a component of G. As T has $s-1$ edges $\Pr[G|_S \cong T] \sim c_T p^{s-1}$, where $c_T = s!/|\mathrm{Aut}(T)|$ is the number of ways T can appear on S. Given $G|_S \cong T$ the probability of S being isolated is $(1-p)^{s(n-s)} \sim e^{-pns} = 1+o(1)$ since $p \ll 1$. Thus $\Pr[A_S] \sim c_T p^{s-1}$. Let X be the number of components of G isomorphic to T so that $X = \sum X_S$ where X_S is the indicator random variable of A_S and

$$E[X] = \sum E[X_S] \sim \binom{n}{s} c_T p^{s-1} \sim c' n^s p^{s-1} \gg 1$$

since $s \le k+1$ and $p \gg n^{-1-1/k}$. We want $\Pr[X \le r] \ll 1$ for all fixed r. This requires a second moment method argument on X which we omit.

What are the countable graphs G satisfying B, C and the schema $A_{T,r}$? By B the components of G must be limited to $k+1$ points and by C these components must be trees. By the schema $A_{T,r}$ every such tree must appear as a component infinitely, i.e., $\geq r$ for all r, often, hence countably often. We have uniquely defined G as consisting of a countable number of components of each such T and nothing else. All countable models of B, C, $A_{T,r}$ are isomorphic. Hence the theory B, C, $A_{T,r}$ is complete. Hence $G_{n,p}$ satisfies the Zero-One Law. □

What happens after $p = n^{-1}$? The example "there exists a K_4" and many others indicated that threshold functions always had a *rational* power of n, possibly times lower order (e.g., logarithmic) terms. In 1988 Saharon Shelah and this author were able to make this notion precise.

THEOREM. Let $0 < \alpha < 1$ be irrational. Set $p = n^{-\alpha}$. Then for any first order sentence A

$$\lim_{n\to\infty} \Pr[G_{n,p} \text{ has } A] = 0 \text{ or } 1.$$

Evolution near $p = 1/n$. The most interesting stage in the evolution of the random graph is near $p = 1/n$. Suppose $p = c/n$ with $c < 1$. The degree of a point has distribution $B(n-1, p)$, which is roughly Poisson c. We estimate the component containing a given point P by a branching process. The point has Poisson c neighbors, or children. Each neighbor has Poisson c neighbors, or grandchildren of P. When $c < 1$ the branching process eventually stops. That is, let p_k be the probability that in the pure branching process there are precisely k descendants, including the original point. With $c < 1$, $\sum_{k=1}^{\infty} p_k = 1$. In the random graph roughly $p_k n$ points will lie in components of size k. The largest component has size $O(\ln n)$, that k for which $p_k + p_{k+1} + \cdots \sim 1/n$.

Now suppose $c > 1$. Then

$$\sum_{k=1}^{\infty} p_k = 1 - \alpha_c$$

where $\alpha_c > 0$ is the probability that the pure branching process goes on forever. Again for each fixed k there are roughly $p_k n$ points in components of size k. The branching process analogy becomes inaccurate for k large as points begin to have common neighbors. What actually happens is that the $\alpha_c n$ points for which the process "lasted forever" lie in a single "giant component." All other components have size at most $O(\ln n)$.

What happens at $c = 1$? The situation is most delicate, as the components of size $O(\ln n)$ are merging to create a giant component. Exactly at $p = 1/n$ the largest component has size $\sim n^{2/3}$ and there are many such components. The large components are most unstable. They tend to merge quickly with each other, as the probability of two components merging is proportional to the product of their cardinalities. The change in the size of the largest component from $O(\ln n)$ to $O(n^{2/3})$ to $O(n)$ is called the Double Jump. I do not feel this term is particularly appropriate as the middle value $O(n^{2/3})$ is found only by choosing $p = 1/n$ precisely.

In recent years the Double Jump has become better understood. The key is the appropriate parametrization to "slow down" the Double Jump. This turns out to be

$$p = \frac{1}{n} + \frac{\lambda}{n^{4/3}}.$$

In computer experiments with $n = 50000$ the transition from, say, $\lambda = -4$ to $\lambda = +4$ is striking. At $\lambda = -4$ the largest component has size $\varepsilon n^{2/3}$ for a small ε and, more important, the top several components are all roughly the same size and they are all trees. When we add $(\Delta\lambda)n^{-4/3}$ to p, components of size $\varepsilon_1 n^{2/3}$ and $\varepsilon_2 n^{2/3}$ have probability $\sim \varepsilon_1\varepsilon_2(\Delta\lambda)$ of merging. By the time $\lambda = +4$ most all of these components have merged and formed a *dominant component* of size $Kn^{2/3}$ for a fairly large K. All the other components are small, less than $\varepsilon' n^{2/3}$ for some small ε'. The dominant component is not a tree and has several more edges than vertices. From then on the dominant component continues to get larger, absorbing the smaller components. While occasionally two small components merge they never challenge the dominant component in terms of relative size. These computer results are supported by a wealth of quite detailed theoretical analysis.

Connectivity. At $p = 100/n$, say, there is a giant component and many small components. As further edges are added the smaller components are sucked up into the giant component until only a few remain. The last surviving small component is an isolated point; when it is joined the graph becomes connected. Erdös and Rényi found a surprisingly precise description of when this occurs.

THEOREM. *Let* $p = p(n) = (\ln n)/n + c/n$. *Then*

$$\lim_{n\to\infty} \Pr[G_{n,p} \textit{ is connected}] = e^{-e^{-c}}.$$

Proof. Let A_i be the event "i is isolated," X_i the associated indicator random variable and $X = X_1 + \cdots + X_n$ the number of isolated points. Set $\mu = \Pr[A_i] = (1-p)^{n-1}$ so that $\mu \sim e^{-pn} = e^{-c}/n$. Then

$$E[X] = n\mu \sim e^{-c}.$$

Let us show X is asymptotically Poisson. According to our earlier notation

$$F^{(r)} = \sum E[X_{i_1} \cdots X_{i_r}] = \binom{n}{r} E[X_1 \cdots X_r]$$

by symmetry. But we can calculate precisely

$$E[X_1 \cdots X_r] = \Pr[1, \cdots, r \text{ isolated}] = (1-p)^{r(n-1)-\binom{r}{2}}$$
$$= \mu^r (1-p)^{-\binom{r}{2}}.$$

For any fixed r, $\lim (1-p)^{-\binom{r}{2}} = 1$ so $F(r) \to \binom{n}{r}\mu^r \to (e^{-c})^r/r!$ The conditions apply, X is asymptotically Poisson and

$$\lim_{n\to\infty} \Pr[X=0] = e^{-e^{-c}}.$$

If $X > 0$, then G is disconnected. If $X = 0$, G may still be disconnected, but we now show that this occurs with probability approaching zero. If $X = 0$ and G is disconnected there is a component of size t for some t, $2 \le t \le n/2$. Let us examine $t = 2$ in detail. There are $\binom{n}{2} < n^2/2$ two-sets. The probability of a given two-set forming a component is $p(1-p)^{2(n-2)}$ as the points must be joined to each other and to nothing else. The expected number of two-point components is then at most $(n^2/2)p(1-p)^{2(n-2)} \sim (p/2)(ne^{-pn})^2 = (p/2)e^{-2c}$, which approaches zero since $p = o(1)$. Similarly, one can show that the expected number of components of all sizes t, $2 \le t \le n/2$ approaches zero. The probability that G is connected is $o(1)$ away from the probability that G has no isolated points and thus is $o(1) + \exp[-e^{-c}]$. $\square$

Addendum: Number theory. The second moment method is an effective tool in number theory. Let $v(n)$ denote the number of primes p dividing n. (We do not count multiplicity though it would make little difference.) The following result says, roughly, that "almost all" n have "very close to" $\ln \ln n$ prime factors. This was first shown by Hardy and Ramanujan in 1920 by a quite complicated argument. We give the proof found by Paul Turan in 1934, a proof that played a key role in the development of probabilistic methods in number theory.

THEOREM. *Let $f(n)$ approach infinity arbitrarily slowly. The number of x in $1, \cdots, n$ such that*

$$|v(x) - \ln \ln n| > f(n)(\ln \ln n)^{1/2}$$

is $o(n)$.

Proof. Let x be randomly chosen from $1, \cdots, n$. For p prime set

$$X_p = \begin{cases} 1 & \text{if } p \mid x, \\ 0 & \text{if not,} \end{cases}$$

and set $X = \sum X_p$, the summation over all primes $p \le n$, so that $X(x) = v(x)$. Now

$$E[X_p] = [n/p]/n.$$

As $y - 1 < [y] \le y$

$$E[X_p] = 1/p + O(1/n).$$

By linearity of expectation

$$E[X] = \sum_{p \le n} (1/p + O(1/n)) = \ln \ln n + o(1).$$

Now we bound the variance:

$$\operatorname{var}(X) = \sum_p \operatorname{var}(X_p) + \sum_{p \ne q} \operatorname{cov}(X_p, X_q).$$

As $\operatorname{var}(X_p) < E(X_p)$ the first sum is at most $\ln \ln n + o(1)$. For the covariance note that $X_p X_q = 1$ if and only if $pq \mid n$. Hence

$$\begin{aligned} \operatorname{cov}(X_p, X_q) &= E(X_p X_q) - E(X_p)E(X_q) \\ &= [n/pq]/n - ([n/p]/n)([n/q]/n) \\ &\le 1/pq - (1/p - 1/n)(1/q - 1/n) \\ &\le 1/n(1/p + 1/q). \end{aligned}$$

Thus

$$\sum_{p\neq q} \operatorname{cov}(X_p, X_q) \leq (1/n) \sum_{p\neq q} (1/p + 1/q)$$
$$= (\pi(n)-1)/n \sum_p 1/p$$

where $\pi(n) \sim n/\ln n$ is the number of primes $p \leq n$. So

$$\operatorname{cov}(X_p, X_q) < \frac{(n/\ln n)}{n} (\ln \ln n) = o(1).$$

That is, the covariances do not affect the variance, $\operatorname{var}(X) = \ln \ln n + o(1)$, and the second moment method, Chebyshev's inequality, actually gives

$$\Pr[|v(x) - \ln \ln n| > K(\ln \ln n)^{1/2}] < K^{-2} + o(1)$$

for any constant K.

REFERENCES

There are two general books on the topic of random graphs:

B. Bollobás, *Random Graphs*, Academic Press, London, 1985.

E. Palmer, *Graphical Evolution*, John Wiley, New York, 1985.

The original seminal paper on the topic is well worth reading:

P. Erdös and A. Rényi, *On the evolution of random graphs*, Magyar Tud. Akad. Mat. Kut. Int. Kozl., 5 (1960), pp. 17–61.

The Zero-One Law for p constant is in:

R. Fagin, *Probabilities on finite models*, J. Symbolic Logic, 41 (1976), pp. 50–58.

The Zero-One Law for $p = n^{-\alpha}$ is in:

S. Shelah and J. Spencer, *Zero-one laws for sparse random graphs*, J. Amer. Math. Soc., 1 (1988), pp. 97–115.

A different, perhaps easier, approach is in:

J. Spencer, *Threshold spectra via the Ehrenfeucht game*, Discrete Appl. Math., 30 (1991), pp. 235–252.

LECTURE 4

Large Deviations and Nonprobabilistic Algorithms

Large deviation. Let us start with a beautiful proof of a result used in Lecture 1.

THEOREM. *Let* $S_n = X_1 + \cdots + X_n$ *where*

$$\Pr[X_i = +1] = \Pr[X_i = -1] = \tfrac{1}{2}$$

and the X_i *are mutually independent. Then for any* $\lambda > 0$

$$\Pr[S_n > \lambda] < e^{-\lambda^2/2n}.$$

Proof. For any $\alpha > 0$

$$E[e^{\alpha X_i}] = \tfrac{1}{2}[e^{\alpha} + e^{-\alpha}] = \cosh(\alpha) \le e^{\alpha^2/2}.$$

(The inequality can be shown by comparing the Taylor Series.) As the X_i are mutually independent,

$$E[e^{\alpha S_n}] = E\left[\prod_{i=1}^{n} e^{\alpha X_i}\right] = \prod_{i=1}^{n} E[e^{\alpha X_i}] < \prod_{i=1}^{n} e^{\alpha^2/2} = e^{\alpha^2 n/2}.$$

Thus

$$\Pr[S_n > \lambda] = \Pr[e^{\alpha S_n} > e^{\alpha\lambda}] \le E[e^{\alpha S_n}]\, e^{-\alpha\lambda} < e^{\alpha^2 n/2 - \alpha\lambda}.$$

We now choose $\alpha = \lambda/n$, optimizing the inequality

$$\Pr[S_n > \lambda] < e^{-\lambda^2/2n}. \qquad \square$$

More generally, let $Y_1, \cdots, Y_n$ be independent with

$$\Pr[Y_i = 1] = p_i, \qquad \Pr[Y_i = 0] = 1 - p_i$$

and normalize by setting $X_i = Y_i - p_i$. Set $p = (p_1 + \cdots + p_n)/n$ and $X = X_1 + \cdots + X_n$. We give the following without proof:

$$\Pr[X > a] < e^{-2a^2/n},$$

$$\Pr[X < -a] < e^{-a^2/2pn},$$

$$\Pr[X > a] < e^{-a^2/2pn + a^3/2(pn)^3}.$$

The last two bounds are useful when $p \ll 1$. The $a^3/2(pn)^3$ term is usually small in applications. When all $p_i = p$, $X = B(n, p) - np$ is roughly normal with zero mean and variance $np(1-p) \sim np$, explaining somewhat the latter two inequalities.

Discrepancy. Let $\mathcal{A} \subseteq 2^\Omega$ be an arbitrary family of finite sets. Let $\chi : \Omega \to \{+1, -1\}$ be a two-coloring of the underlying points. Define

$$\chi(A) = \sum_{a \in A} \chi(a),$$

$$\text{disc}\,(\chi) = \max_{A \in \mathcal{A}} |\chi(A)|,$$

$$\text{disc}\,(\mathcal{A}) = \min_{\chi} \text{disc}\,(\chi).$$

Note that disc $(\chi) \leq K$ means that there is a two-coloring of Ω so that every $A \in \mathcal{A}$ has $|\chi(A)| \leq K$.

THEOREM. *If* $|\mathcal{A}| = |\Omega| = n$, *then*

$$\text{disc}\,(\mathcal{A}) \leq \sqrt{2n \ln (2n)}.$$

Proof. With χ random and $|A| = r$, $\chi(A)$ has distribution S_r. As all $A \subseteq \Omega$ have $|A| \leq |\Omega| = n$

$$\Pr [|\chi(A)| > \lambda] < 2\, e^{-\lambda^2/2n}.$$

Thus

$$\Pr [\text{disc}\,(\chi) > \lambda] < \sum_{A \in \mathcal{A}} \Pr [|\chi(A)| > \lambda] < 2n\, e^{-\lambda^2/2n} = 1$$

by taking $\lambda = \sqrt{2n \ln (2n)}$. Thus $\Pr [\text{disc}\,(\chi) \leq \lambda] > 0$ and so there exists a χ with disc $(\chi) \leq \lambda$.

We can also express this result in vector form.

THEOREM. *Let* $u_j \in R^n$, $|u_j|_\infty \leq 1$, $1 \leq j \leq n$. *Then there exist* $\varepsilon_j \in \{-1, +1\}$ *so that, setting* $u = \varepsilon_1 u_1 + \cdots + \varepsilon_n u_n$, $|u|_\infty \leq \sqrt{2n \ln (2n)}$. (Note: With $u = (L_1, \cdots, L_n)$, $|u|_\infty = \max |L_i|$, the L^∞ norm.)

Here is the translation between the formulations. Given $\mathcal{A} \subseteq 2^\Omega$, number the elements $1, \cdots, n$ and the sets $A_1, \cdots, A_n$ and define the incidence matrix $A = [a_{ij}]$ by $a_{ij} = 1$ if $j \in A_i$; 0 otherwise. Let u_j be the column vectors. A two-coloring $\chi : \Omega \to \{+1, -1\}$ corresponds to $\varepsilon_j = \chi(j)$, disc (A_i) to the ith coordinate L_i of u and disc (χ) to $|u|_\infty$. Now, however, we allow $a_{ij} \in [-1, +1]$ to be arbitrary. The proof of our first theorem can be easily modified to show that with ε_j random $\Pr [|L_i| > \lambda] < 2\, e^{-\lambda^2/2n}$ and the rest of the proof follows as before.

$$\begin{array}{c} \\ S_1 \\ \vdots \\ S_n \\ \\ \\ \end{array} \begin{array}{c} 1 \quad 2 \quad \cdots \quad n \\ \left[\begin{array}{ccc} & & \\ & & \\ & & \end{array} \right] \\ u_1 \qquad\qquad u_n \\ + \quad - \qquad\quad - \end{array} \qquad \begin{array}{l} L_{11} = \chi(S_1) \\ \quad\vdots \\ L_{1n} = \chi(S_n) \\ u = \pm u_1 \pm \cdots \pm u_n = (L_{11}, \cdots, L_{1n}) \end{array}$$

A dynamic algorithm. We have a probabilistic proof that the ε_j exist with $|u|_\infty < \lambda = \sqrt{2n \ln(2n)}$. Now we seek an algorithm which finds such ε_j. Of course, the algorithms should work in polynomial time; we cannot simply check all 2^n choices of the ε_j. If we modify our requirements slightly and ask for, say, $|u|_\infty \le \sqrt{2n \ln(2n)}$ (1.1) then a random algorithm is immediate. Simply pick the ε_j at random. The failure probability is at most $2n(2n)^{-1.21} \ll 1$. But now we do not allow a probabilistic step.

Here is a very general method to "remove the coin flip" that has only one flaw: Suppose the set of all n-tuples $(\varepsilon_1, \cdots, \varepsilon_n)$, $\varepsilon_i \in \{0, 1\}$, each equiprobable is our underlying probability space and $A_1, \cdots, A_m$, are "bad" events with $\sum \Pr[A_i] < 1$. We want to find $\varepsilon_1, \cdots, \varepsilon_n$ so that $\bigwedge \bar{A}_i$ is satisfied. (In our case $m = n$ and A_i is the event $|L_i| \ge \lambda$.) We find $\varepsilon_1, \varepsilon_2, \cdots$ sequentially. Assume $\varepsilon_1, \cdots, \varepsilon_{j-1}$ have already been fixed. Set

$$W_i = \Pr[A_i | \varepsilon_1, \cdots, \varepsilon_{j-1}].$$

That is, W_i is the probability that A_i will occur with $\varepsilon_1, \cdots, \varepsilon_{j-1}$ already fixed and the remaining ε_k to be chosen by independent coin flips. Fixing ε_j changes the W_i in two possible ways. Set

$$W_i^+ = \Pr[A_i | \varepsilon_1, \cdots, \varepsilon_{j-1}, \varepsilon_j = +1],$$

$$W_i^- = \Pr[A_i | \varepsilon_1, \cdots, \varepsilon_{j-1}, \varepsilon_j = -1].$$

Then $W_i = \frac{1}{2}[W_i^+ + W_i^-]$. Now set $W^{\text{OLD}} = \sum W_i$, $W^{\text{OLD}+} = \sum W_i^+$, $W^{\text{OLD}-} = \sum W_i^-$ so that $W^{\text{OLD}} = \frac{1}{2}[W^{\text{OLD}+} + W^{\text{OLD}-}]$. Here is the algorithm: If $W^{\text{OLD}+} \le W^{\text{OLD}-}$ select $\varepsilon_j = +1$; otherwise select $\varepsilon_j = -1$.

Having chosen ε_j let

$$W^{\text{NEW}} = \sum \Pr[A_i | \varepsilon_1, \cdots, \varepsilon_{j-1}, \varepsilon_j].$$

(That is, we have selected ε_j so as to minimize W^{NEW}.) Then

$$W^{\text{NEW}} = \min[W^{\text{OLD}+}, W^{\text{OLD}-}] \le \tfrac{1}{2}[W^{\text{OLD}+} + W^{\text{OLD}-}] = W^{\text{OLD}}.$$

Let W^{INIT}, W^{FIN} denote the values of W at $j = 1$ (before any of the ε_j have been selected) and $j = n+1$ (after all of the ε_j have been selected), respectively. As $W^{\text{NEW}} \le W^{\text{OLD}}$ at each stage

$$W^{\text{FIN}} \le W^{\text{INIT}}.$$

But $W^{\text{INIT}} = \sum \Pr[A_j] < 1$ by assumption. Hence $W^{\text{FIN}} < 1$. But now the point $(\varepsilon_1, \cdots, \varepsilon_n)$ of the probability space is fixed so all conditional probabilities are either zero or one. (Either the A_i have occurred or they have not.) The sum is less than one so all the addends must be zero. All of the A_i have not occurred, $(\varepsilon_1, \cdots, \varepsilon_n)$ satisfies $\bigwedge \bar{A}_i$, the algorithm has succeeded.

Where is the flaw? The algorithm always works, yes, but not always in polynomial time. The calculation of the conditional probabilities may take exponential time. In our situation even calculation of $\Pr[A_i]$ requires the number of $(\varepsilon_1, \cdots, \varepsilon_n)$ so that $|\sum \varepsilon_j a_{ij}| > \sqrt{2n \ln(2n)}$ and there is no polynomial algorithm known for this when the $a_{ij} \in [-1, +1]$ are arbitrary. We do have success when all $a_{ij} \in \{0, 1\}$, our original problem. Then all conditional probabilities can be expressed easily in terms of appropriate binomial coefficients and the determination of the ε_j can be done in polynomial time. We will soon see another algorithm that works for arbitrary a_{ij}.

The pusher-chooser game. Let us consider a zero-sum two-player game consisting of n rounds. There is a point $P \in R^n$ called the position vector which will be moved around; it is set initially at 0. On each round the first player, called Pusher, selects $v \in R^n$ with $|v|_\infty \leq 1$ and then (knowing P and v) the second player, called Chooser, resets P to either $P+v$ or $P-v$. Let P^{FIN} denote the value of P after the completion of n rounds. The payoff to Pusher is $|P^{\mathrm{FIN}}|_\infty$. Let VAL (n) be the value (to Pusher) of this game. (The same game with $|v|_2 \leq 1$ and payoff $|P^{\mathrm{FIN}}|_2$

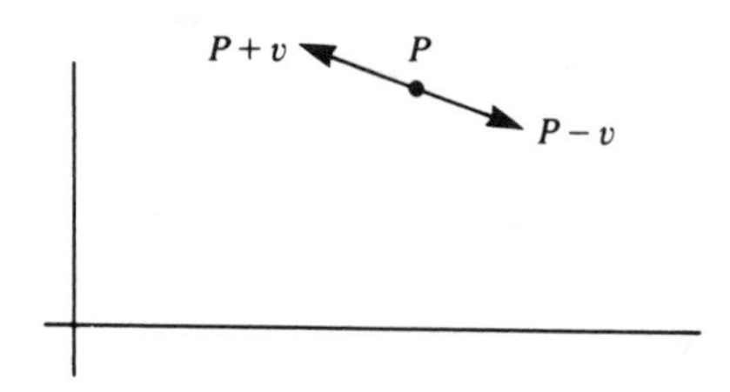

has a simple solution. Pusher can always choose a unit v orthogonal to P so that, regardless of Chooser's choice, $|P^{\mathrm{NEW}}|^2 = |P^{\mathrm{OLD}}|^2 + |v|^2$ and so $|P^{\mathrm{FIN}}| = n^{1/2}$. On the other hand, Chooser may always select a sign so that $\pm v$ makes an acute or right angle with P and with this strategy $|P^{\mathrm{FIN}}| \leq n^{1/2}$. Hence the game has value $n^{1/2}$.)

A hyperbolic cosine algorithm.

THEOREM. VAL $(n) \leq \sqrt{2n \ln (2n)}$.

Proof. Let $\alpha > 0$ be fixed but, for the moment, arbitrary. Set $G(x) = \cosh(\alpha x)$. For any $a \in [-1, +1]$

$$(*) \qquad \tfrac{1}{2}[G(x+a) + G(x-a)] = G(x)G(a) \leq G(x) \cosh(\alpha) \leq G(x)\, e^{\alpha^2/2}.$$

For $x \in R^n$, $x = (x_1, \cdots, x_n)$, set

$$G(x) = \sum_{i=1}^{n} G(x_i).$$

For any $P \in R^n$ and $v \in R^n$ with $|v|_\infty \leq 1$, summing $(*)$ over the coordinates gives

$$(**) \qquad \tfrac{1}{2}[G(P+v) + G(P-v)] \leq G(P)\, e^{\alpha^2/2}.$$

Here is an algorithm for Chooser: At all times select that sign which minimizes the new value of $G(P)$.

Let P^{OLD}, P^{NEW} be the values of P at the beginning and end of any particular move with v the choice of Pusher. Then

$$\begin{aligned} G(P^{\mathrm{NEW}}) &= \min [G(P^{\mathrm{OLD}} + v), G(P^{\mathrm{OLD}} - v)] \\ &\leq \tfrac{1}{2}[G(P^{\mathrm{OLD}} + v) + G(P^{\mathrm{OLD}} - v)] \\ &< G(P^{\mathrm{OLD}})\, e^{\alpha^2/2}. \end{aligned}$$

At the beginning of the game $G(P)=G(0)=n$ so

$$G(P^{\mathrm{FIN}})<n\,e^{\alpha^2 n/2}.$$

For any x, $G(x)\geq \cosh(\alpha|x|_\infty)>\frac{1}{2}e^{\alpha|x|_\infty}$ so

$$\tfrac{1}{2}e^{\alpha|P^{\mathrm{FIN}}|_\infty}\leq n\,e^{\alpha^2/2},\qquad |P^{\mathrm{FIN}}|_\infty\leq \ln(2n)/\alpha+\alpha n/2.$$

Setting $\alpha=\sqrt{2\ln(2n)/n}$ we get the best algorithm.

$$\mathrm{VAL}(n)\leq|P^{\mathrm{FIN}}|_\infty<\sqrt{2n\ln(2n)}.\qquad\square$$

Calculation of $G(P)$ can be done rapidly. This result gives a good algorithm for finding $\varepsilon_1,\cdots,\varepsilon_n=\pm1$ so that $|\sum\varepsilon_j u_j|_\infty<\sqrt{2n\ln(2n)}$. The algorithm is *nonanticipative* in that the decision on ε_j does not require "lookahead" to the data a_{ik} with $k>j$. The number VAL (n) is the least value for $|\sum\varepsilon_j u_j|_\infty$ that can be achieved by a nonanticipative (though not necessarily rapid) algorithm.

Let the chips fall where they may. Now we give a lower bound for VAL (n). Here is a simple strategy for Pusher to decide his move v. Given P, group all coordinates by their value. In each group let half of the v coordinates be $+1$ and half be -1. If there are an odd number in the group let one coordinate be zero. For example,

$$\begin{array}{lllllllllllllllll} \text{If} & P= & 0 & 0 & 0 & 0 & 0 & 1 & 1 & 1 & 1 & 2 & 2 & 2 & 3 & 4 & 4\\ \text{set} & v= & + & + & - & - & 0 & + & + & - & - & + & - & 0 & 0 & + & -. \end{array}$$

With this strategy Chooser has no choice! In the example, after permutation of the coordinates

$$P\pm v=-1\ \ -1\ \ 1\ \ 1\ \ 0\ \ 2\ \ 2\ \ 0\ \ 0\ \ 3\ \ 1\ \ 2\ \ 3\ \ 5\ \ 3.$$

With P initially 0, P^{FIN} is determined and $|P^{\mathrm{FIN}}|_\infty$ gives a lower bound for VAL (n). It only remains to estimate $|P^{\mathrm{FIN}}|_\infty$.

It is more convenient to work with the distribution of the coordinate values. Given $P=(x_1,\cdots,x_n)\in Z^n$ (with this strategy P is always a lattice point) define $\hat{P}:Z\to N$ by letting $\hat{P}(x)$ be the number of i with $x_i=x$. With P in our example we may write

$$\hat{P}=\underline{5}\,4\,3\,1\,2$$

listing the values of $\hat{P}(x)$, underscoring $\hat{P}(0)$. We envision a one-dimensional array of squares with $P(x)$ chips on square x.

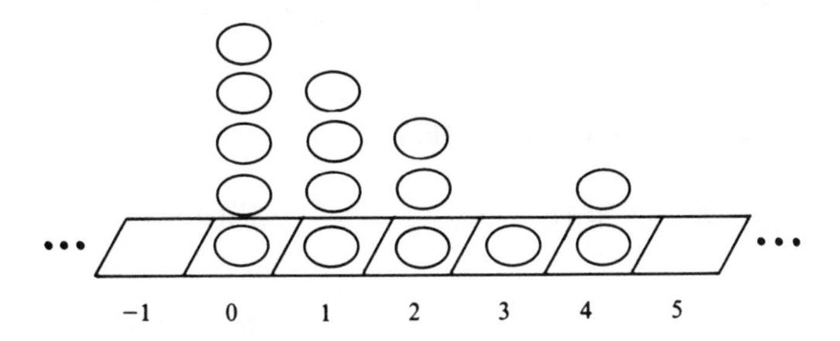

After Pusher and Chooser move, the new position has

$$P^{\text{NEW}} = 2\,\underline{3}\,3\,3\,3\,0\,1.$$

Let T be the transformation taking P^{OLD} to P^{NEW}. In terms of chips (the way we like to think of the problem!) applying T consists of simultaneously taking each pile of chips and moving half of them one square to the left, half of them one square to the right, and leaving the odd one (if the pile is odd) where it is. For a chip position $\hat{P}$ define the support supp $(\hat{P})$ as the maximal i such that there are chips at either $+i$ or $-i$. Let $\bar{n}$ be the position consisting of n chips on square zero. Then

$$\text{VAL}\,(n) \geq \text{supp}\,(T^n(\bar{n})).$$

For example, when $n = 20$ the values of $T^i(\overline{20})$ are $10\,\underline{0}\,10$, $5\,0\,\underline{10}\,0\,5$, $2\,0\,7\,\underline{1}\,7\,0\,2$, $\cdots$, $1\,1\,1\,1\,1\,1\,3\,1\,\underline{2}\,1\,3\,1\,1\,1\,1\,1$ so that VAL $(20) \geq 7$. Incidentally, the chips game is marvelous to study in its own right. All you need is a small computer and a large imagination. Start with a hundred chips on square zero, or whatever, and carry on! There is no need to stop at time one hundred—the death throes of the process are fascinating.

Set $a = \text{supp}\,(T^n(\bar{n}))$. We estimate a by approximating T with a linear transformation S, for which chips may be split in half or further. For somewhat technical reasons set $\Omega = \{-a, \cdots, a\} \cup \{\infty\}$ and define a linear operator S by (here $x : \Omega \to R$)

$$(Sx)(i) = \tfrac{1}{2}[x(i-1) + x(i+1)], \ |i| < a,$$

$$(Sx)(a) = \tfrac{1}{2}x(a-1),$$

$$(Sx)(-a) = \tfrac{1}{2}x(-a+1),$$

$$(Sx)(\infty) = x(\infty) + \tfrac{1}{2}x(a) + \tfrac{1}{2}x(-a).$$

For chip positions in $(-a, +a)$, S acts like T except that odd chips are split. S is the linear operator for the random walk on Ω with an absorbing barrier at ∞. The definition of a allows us to consider $T^n(\bar{n})$ well defined on Ω.

CLAIM. $(S^n - T^n)(\bar{n})(\infty) \leq 2a + 1$.

Proof.

$$(S^n - T^n)(\bar{n})(\infty) = \sum_{i=0}^{n-1} S^i(S-T)T^{n-1-i}(\bar{n})(\infty).$$

Suppose $T^{n-1-i}(\bar{n})$ consists of a_{ij} chips at position j, $-a \leq j \leq a$. S and T act identically on each even pile of chips and differ only on the odd chips, where T acts like the identity. Letting u_j denote a single chip at position j

$$(S-T)T^{n-1-i}(\bar{n}) = \sum_{j \in A_i} Su_j - u_j$$

where A_i is the set of j with a_{ij} odd. Then

$$(S^n - T^n)(\bar{n})(\infty) = \sum_{i=0}^{n-1} \sum_{j \in A_i} (S^{i+1} - S^i)(u_j)(\infty)$$

$$\leq \sum_{i=0}^{n-1} \sum_{|j| \leq a} (S^{i+1} - S^i)(u_j)(\infty)$$

as for all i, j $(S^{i+1}-S^i)(u_j)(\infty)\geq 0$. This is the technical reason we gave S an absorbing barrier at ∞. So

$$\begin{aligned}(S^n-T^n)(\bar{n})(\infty) &\leq \sum_{|j|\leq a}\sum_{i=0}^{n-1}(S^{i+1}-S^i)(u_j)(\infty)\\ &= \sum_{|j|\leq a}(S^n-I)(u_j)(\infty)\\ &\leq \sum_{|j|\leq a}(1-0)=2a+1.\end{aligned}$$ □

What about the asymptotics of a? We have chosen a so that $T^n(\bar{n})(\infty)=0$ and hence $S^n(\bar{n})(\infty)\leq 2a+1$ and so, by linearity

$$(*)\qquad S^n(\bar{1})(\infty)\leq(2a+1)/n.$$

But $S^n(\bar{1})(\infty)$ is simply the probability of reaching ∞ in the random walk on Ω with absorbing barrier at ∞ when you start at zero. This is simply the probability of ever reaching $\pm(a+1)$ with time n in the normal one-dimensional random walk. When $a=Kn^{1/2}$ with K slowly going to infinity this probability is $e^{-K^2/2(1+o(1))}$. For any fixed $\varepsilon>0$ (omitting the calculations) $(*)$ fails if $a=\sqrt{n\ln n}(1-\varepsilon)$ and so

$$\mathrm{VAL}(n)>\sqrt{n\ln n}(1-o(1)).$$

REFERENCES

The Pusher-Chooser game and the Hyperbolic Cosine Algorithm are from:

J. SPENCER, *Balancing games*, J. Combin. Theory Ser. B, 23 (1977), pp. 68–74.

The Chips game is in:

J. SPENCER, *Balancing vectors in the* max *norm*, Combinatorica, 6 (1986), pp. 55–65.

and

R. J. ANDERSON, L. LOVÁSZ, P. SHOR, J. SPENCER, E. TARDOS, S. WINOGRAD, *Disks, balls and walls: Analysis of a Combinatorial game*, Amer. Math. Monthly, 96 (1989), pp. 481–493.

The large deviation results are given in an appendix to *The Probabilistic Method*, referenced in the Preface.

LECTURE 5

Discrepancy I

Hereditary and linear discrepancy. We recall the definition of the discrepancy of a set system $\mathcal{A} \subseteq 2^{\Omega}$

$$\text{disc}(\mathcal{A}) = \min_{\chi} \max_{A \in \mathcal{A}} |\chi(A)|$$

where χ ranges over "two-colorings" $\chi: \Omega \to \{-1, +1\}$ and $\chi(A) = \sum_{a \in A} \chi(a)$. For $X \subset \Omega$ let $\mathcal{A}|_X$ be the restriction of $\mathcal{A}$ to X, the family of sets $A \cap X$, $A \in \mathcal{A}$. We define the *hereditary discrepancy*

$$\text{herdisc}(\mathcal{A}) = \max_{X \subset \Omega} \text{disc}(\mathcal{A}|_X).$$

Let $\Omega = \{1, \cdots, n\}$ for convenience. We define the *linear discrepancy*

$$\text{lindisc}(\mathcal{A}) = \max_{p_1, \cdots, p_n \in [0,1]} \min_{\varepsilon_1, \cdots, \varepsilon_n \in \{0,1\}} \max_{A \in \mathcal{A}} \left| \sum_{i \in A} (\varepsilon_i - p_i) \right|.$$

We think of $p_1, \cdots, p_n$ as "initial values" and $\varepsilon_1, \cdots, \varepsilon_n$ as a "simultaneous approximation." Then $E_A = \sum_{i \in A} \varepsilon_i - p_i$ represents the error in the approximation with respect to a set A. In this context $\text{lindisc}(\mathcal{A}) \leq K$ means that given any initial values there is a simultaneous approximation so that $|E_A| \leq K$ for all $A \in \mathcal{A}$. When all $p_i = \frac{1}{2}$ the possible "errors" $\varepsilon_i - p_i = \pm\frac{1}{2}$ and the problem reduces to discrepancy. Thus

$$\text{lindisc}(\mathcal{A}) \geq \tfrac{1}{2} \text{disc}(\mathcal{A}).$$

But an upper bound on lindisc is much stronger. For example, $\text{lindisc}(\mathcal{A}) \leq K$ implies, taking all $p_i = \frac{1}{3}$, that there is a set $S = \{i: \varepsilon_i = 1\}$ so that

$$E_A = |A \cap S| - |A|/3$$

satisfies $|E_A| \leq K$ for all $A \in \mathcal{A}$. Finally we define the hereditary linear discrepancy

$$\text{herlindisc}(\mathcal{A}) = \max_{X \subseteq \Omega} \text{lindisc}(\mathcal{A}|_X).$$

Some examples. (1) Let $\Omega = [n]$ and let $\mathcal{A}$ consist of all singletons and $[n]$ itself. Then $\text{disc}(\mathcal{A}) = 1$, as we may color half the points "Red" $(= +1)$ and half "Blue." Also $\text{herdisc}(\mathcal{A}) = 1$ as $\mathcal{A}|_X$ has the same form as $\mathcal{A}$ for all $X \subset \Omega$. When all $p_i = \frac{1}{2}$ we may set half the $\varepsilon_i = +1$, half -1 so that all $|E_A| \leq \frac{1}{2}$. But these are not the "worst" initial values. Set all $p_i=1/(n+1)$. If any $\varepsilon_i=+1$, then $E_{\{i\}}=n/(n+1)$ whereas if all $\varepsilon_i=0$, then $E_\Omega=-n/(n+1)$. Thus lindisc $(\mathcal{A})\geq n/(n+1)$ and in fact (exercise) equality holds.

(2) Let $\Omega=[2n]$, n even, and let $\mathcal{A}$ consist of all $S\subset\Omega$ with $|S\cap\{1,\cdots,n\}|=|S|/2$. Then disc $(\mathcal{A})=0$ by setting $\varepsilon_i=+1$, $1\leq i\leq n$, $\varepsilon_i=-1$, $n<i\leq 2n$. Let $X=\{1,\cdots,n\}$ so that $\mathcal{A}|_X=2^X$. disc $(\mathcal{A}|_X)=n/2$ since given $\varepsilon_1,\cdots,\varepsilon_n=\pm 1$ we may examine either $\{i\colon \varepsilon_i=+1\}$ or its complement. Thus

$$\text{herdisc}\,(\mathcal{A})\geq \text{disc}\,(\mathcal{A}|_X)=n/2.$$

A lower bound on lindisc $(\mathcal{A})$ can be given by setting $p_i=\frac{1}{2}$, $1\leq i\leq n$, $p_i=0$ elsewhere. Let $\varepsilon_1,\cdots,\varepsilon_{2n}\in\{0,1\}$. We claim $|E_A|\geq n/4$ for some $A\in\mathcal{A}$. Basically, we can let A consist of $n/2$ indices from the first half with equal ε_i and $n/2$ indices from the second half with equal ε_i.

The precise values of herdisc $(\mathcal{A})$ and lindisc $(\mathcal{A})$ are interesting problems. I feel that disc $(\mathcal{A})$ may be small "by accident," whereas herdisc $(\mathcal{A})$ and lindisc $(\mathcal{A})$ give more intrinsic properties of the set-system $\mathcal{A}$. To make an analogy, a graph G may "accidentally" have $\omega(G)=\chi(G)$, but if $\omega(H)=\chi(H)$ for all subgraphs H of G, then G is perfect.

(3) Let $\Omega=[n]$ and let $\mathcal{A}$ consist of all intervals $[i,j]$, $1\leq i\leq j\leq n$. Now disc $(\mathcal{A})\leq 1$ since we may alternatively color $[n]$ by $\varepsilon_i=(-1)^i$. For any $X\subset\Omega$

R	B	R	B	R	B	R	B	coloring for [8]
1	2	3	4	5	6	7	8	
R	B		R		B		R	coloring for 1, 2, 4, 6, 8

$\mathcal{A}|_X$ has the same form as $\mathcal{A}$ and disc $(\mathcal{A}|_X)=1$. That is, if X has elements $x_1<x_2<\cdots<x_s$ set $\varepsilon_{x_i}=(-1)^i$. Thus herdisc $(\mathcal{A})=1$. What about lindisc $(\mathcal{A})$? Clearly lindisc $(\mathcal{A})\geq n/(n+1)$ as $\mathcal{A}$ contains the sets of example 1. Here is a gem, shown to me by László Lovász, that proves equality.

THEOREM. *For all $p_1,\cdots,p_n\in[0,1]$ there exist $\varepsilon_1,\cdots,\varepsilon_n\in\{0,1\}$ so that*

$$\left|\sum_{k=i}^{j}\varepsilon_k-p_k\right|\leq n/(n+1)$$

for all $1<i<j<n$.

We outline the argument. Set $q_0=0$, $q_i=p_1+\cdots+p_i$. Consider $q_i\in R/Z$. For any α we may define ε_i to be one if the interval $(q_{i-1},q_i]$ includes α, otherwise zero. Then $p_i+\cdots+p_j$ is the "distance traveled" from q_{i-1} to q_j and $\varepsilon_i+\cdots+\varepsilon_j$ is the number of times α was passed so these numbers differ by at most one. The points $q_0,\cdots,q_n$ split R/Z into $n+1$ intervals so one of these intervals has length at least $1/(n+1)$. The precise result is obtained when α is chosen so that $(\alpha-1/(n+1),\alpha)$ contains none of the q_i.

Matrix form. We can extend our notions of discrepancy by defining for any $m\times n$ matrix H

$$\text{disc}\,(H)=\min_{x\in\{-1,+1\}^n}|Hx|_\infty,$$

herdisc $(H) = \max \text{disc}(H')$, H' a submatrix of H consisting of a subset of the columns,

$$\text{lindisc}(H) = \max_{p\in[0,1]^n} \max_{X\in\{0,1\}^n} |H(p-x)|_\infty.$$

When H is the incidence matrix for a set system $\mathcal{A}$ these definitions reduce to those previously given. All results we show for set systems apply equally well to matrices with all coordinates in $[-1, +1]$. Whether this is always the case or whether set systems have some special properties remains an elusive mystery.

Reduction by linear algebra. When $\mathcal{A}$ has more points than sets we can "reduce" the number of points by linear algebra. Let

$$S^{n,k} = \{(x_1, \cdots, x_n) \in [0,1]^n \colon x_i = 0 \text{ or } 1 \text{ for all but at most } k \text{ } i\text{'s}\}$$

denote the "k-dimensional skeleton" of $[0,1]^n$.

THEOREM. *Let H be an $m \times n$ matrix, $p = (p_1, \cdots, p_n) \in [0,1]^n$. Then there exists $x \in S^{n,m}$ with $Hx = Hp$.*

Proof. If $p \in S^{n,m}$ set $x = p$. Otherwise consider $Hx = Hp$ as a system of equations in variables x_i where if $p_i = 0$ or 1 we consider $x_i = p_i$ as a constant. As there are more variables than equations, there is a line of solutions $x = p + \lambda y$. (This y may be found in polynomial time by standard linear algebraic techniques.) Let λ be the minimal positive real number so that $x_i = p_i + \lambda y_i = 0$ or 1 for some i with $p_i \neq 0, 1$. (Physically, we move along a line of the hyperplane $Hx = Hp$ until we hit the boundary.) For this λ $Hx = Hp$, $x \in [0,1]^n$ and x has at least one more coefficient 0 or 1 than p. Now replace p with x and iterate this procedure. In at most n-m iterates we find x with $Hx = Hp$ and $x \in S^{n,m}$. □

We may also think topologically that $W = \{x \in R^n \colon Hx = Hp\}$ is an affine space of codimension at most m containing a point of $[0,1]^n$; there is no way for W to escape without intersecting the skeleton $S^{n,m}$. For example, in three dimensions a line must intersect some face of $[0,1]^3$ and a plane must intersect some edge.

THEOREM. *Let $\mathcal{A} \subseteq 2^\Omega$ with $|\mathcal{A}| = m$, $|\Omega| = n$. Assume* lindisc $(\mathcal{A}|_X) \leq K$ *for all $X \subset \Omega$ with at most m elements. Then* lindisc $(\mathcal{A}) \leq K$.

Proof. Let H be the incidence matrix of $\mathcal{A}$ and let $p = (p_1, \cdots, p_n) \in [0,1]^n$ be given. The above procedure gives $x = (x_1, \cdots, x_n) \in S^{n,m}$ with $Hx = Hp$. That is,

$$0 = \sum_{i\in A} (x_i - p_i) \quad \text{for all } A \in \mathcal{A}.$$

Let $X = \{i \colon x_i \neq 0, 1\}$. By definition there exist $y_i = 0, 1$ for $i \in X$ so that

$$\left|\sum_{i\in A} (y_i - x_i)\right| \leq \text{lindisc}(\mathcal{A}|_X) \leq K \quad \text{for all } A \in \mathcal{A}.$$

For $i \notin X$ set $y_i = x_i$. Then

$$\left|\sum_{i\in A} (y_i - p_i)\right| = \left|\sum_{i\in A} (y_i - x_i) + \sum_{i\in A} (x_i - p_i)\right|$$

$$= \left|\sum_{i\in A} (y_i - x_i)\right| \leq \text{lindisc}(\mathcal{A}|_X) \leq K \quad \text{for all } A \in \mathcal{A}. \qquad \square$$

Reduction to approximating $\frac{1}{2}$. The next result "reduces" the problem of linear discrepancy—approximating arbitrary $p_1, \cdots, p_n$—to the case when all $p_i = \frac{1}{2}$.

THEOREM. lindisc $(\mathscr{A}) \leq$ herdisc $(\mathscr{A})$.

Proof. Let $K =$ herdisc $(\mathscr{A})$ and fix initial values $p_1, \cdots, p_n$. Assume that all p_i have finite binary expressions and let T be the minimal integer so that all $p_i 2^T$ are integral. Let X be the set of i such that p_i has a "one" in its Tth digit. As disc $(\mathscr{A}|_X) \leq$ herdisc $(\mathscr{A}) = K$ there exist $\varepsilon_i = \pm 1$, $i \in X$ so that

$$\left| \sum_{i \in A \cap X} \varepsilon_i \right| \leq K, \qquad A \in \mathscr{A}.$$

We round off p_i to new values p_i' given by

$$p_i' = \begin{cases} p_i & \text{if } i \notin X, \\ p_i + 2^{-T}\varepsilon_i & \text{if } i \in X. \end{cases}$$

We have created an "error":

$$\left| \sum_{i \in A} (p_i' - p_i) \right| = 2^{-T} \left| \sum_{i \in A \cap X} \varepsilon_i \right| \leq 2^{-T} K$$

for every $A \in \mathscr{A}$ and the new p_i' have binary expressions of length at most $T-1$. Iterate this procedure T times until all values are either 0 or 1. Let $p_i^{(s)}$ denote the values in this procedure when binary expressions have length at most s. Then $p_i^{(T)}$ are the initial p_i and $p_i^{(0)}$ are the final 0 or 1 and

$$\begin{aligned} \left| \sum_{i \in A} p_i^{(0)} - p_i^{(T)} \right| &\leq \sum_{s=1}^{T} \left| \sum_{i \in A} p_i^{(s-1)} - p_i^{(s)} \right| \\ &\leq \sum_{s=1}^{T} 2^{-s} K \\ &\leq \sum_{s=1}^{\infty} 2^{-s} K = K \end{aligned}$$

as desired. □

"Wait a minute!" cries the critic. "What if the p_i do not all have finite binary expressions?" The reply depends on one's orientation. The mathematician replies, "Use compactness." The computer scientist says the question is meaningless—all numbers have finite binary expressions.

COROLLARY. *Let* $|\mathscr{A}| = m \leq n = |\Omega|$ *and suppose* disc $(\mathscr{A}|_Y) \leq K$ *for all* $Y \subset \Omega$ *with at most* m *points. Then* disc $(\mathscr{A}) \leq 2K$.

Proof.

$$\begin{aligned} \text{disc}(\mathscr{A}) &\leq 2 \,\text{lindisc}(\mathscr{A}) \\ &\leq 2 \max_{|X| \leq m} \text{lindisc}(\mathscr{A}|_X) \\ &\leq 2 \max_{|X| \leq m} \text{herdisc}(\mathscr{A}|_X) \\ &\leq 2 \max_{|Y| \leq m} \text{disc}(\mathscr{A}|_Y) \leq 2K. \end{aligned}$$

□

Usually, theorems about discrepancy are actually theorems about hereditary discrepancy. This corollary then allows a reduction to the case "sets = points." For example, in Lecture 4 we showed that any family of n sets on n points has discrepancy at most $[2n \ln (2n)]^{1/2}$.

COROLLARY. *If $\mathcal{A}$ consists of n sets of arbitrary size*

$$\operatorname{disc}(\mathcal{A}) \le 2[2n \ln (2n)]^{1/2}.$$

Straight probabilistic methods could not achieve this result. If a set has size x then a random coloring gives discrepancy about $x^{1/2}$ which can be arbitrarily large. Combining linear algebra with the probabilistic method is very powerful. Note that all the steps are algorithmic. There is a polynomial (in m, n) time algorithm that gives the desired coloring.

Open Problem. What is the maximal c_n so that if $|\Omega| = n$

$$\operatorname{lindisc}(\mathcal{A}) \le (1 - c_n) \operatorname{herdisc}(\mathcal{A})?$$

To show $c_n > 0$ consider the proof of $\operatorname{lindisc}(\mathcal{A}) \le \operatorname{herdisc}(\mathcal{A})$ as an algorithm and let X_s be the set of i such that $p_i^{(s)}$ has a "one" in the sth binary digit. Each $X_s \subset [n]$ so at some time the X_s must repeat. This allows an improvement. In particular there must be $1 \le s < t \le 2^n + 1$ with $X_s = X_t$. We modify the algorithm at "stage s" and set

$$p_i^{(s-1)} = p_i^{(s)} - 2^{-s}\varepsilon_i \quad \text{for } i \in X_s$$

keeping all else the same. For any $A \in \mathcal{A}$

$$\sum_{i \in A} (p_i^{(s-1)} - p_i^{(s)}) = -2^{-s} \sum_{i \in A \cap X} \varepsilon_i,$$

$$\sum_{i \in A} (p_i^{(t-1)} - p_i^{(t)}) = +2^{-t} \sum_{i \in A \cap X} \varepsilon_i.$$

The two "errors" have opposite signs and cancel each other somewhat.

$$\left| \sum_{i \in A} p_i^{(0)} - p_{(i)}^{T} \right| \le \sum_{\substack{u=1 \\ u \ne s,t}}^{T} \left| \sum_{i \in A} p_i^{(u-1)} - p_i^{(u)} \right| + \left| \sum_{i \in A} p_i^{(s-1)} - p_i^{(s)} + p_i^{(t-1)} - p_i^{(t)} \right|$$

$$\le K \sum_{u \ne s,t} 2^{-u} + K(2^{-s} - 2^{-t}) \le K(1 - 2^{-t+1}).$$

As $t \le 2^n + 1$ we have shown that $c_n \ge 2^{-2^n}$. This may well be extremely weak, since example 1 gives the best upper bound to c_n that we know: $c_n \le 1/(n+1)$. There is clearly room for improvement!

Simultaneous round-off. Let data a_{ij}, $1 \le i \le m$, $1 \le j \le n$ be given, all $a_{ij} \in [-1, +1]$. Let initial real values x_j, $1 \le j \le n$ be given. We ask for a simultaneous round-off y_j, each y_j being either the integer "round-up" or "round-down" of x_j. Let E_i be the "error"

$$E_i = \sum_{j=1}^{n} a_{ij} x_j - \sum_{j=1}^{n} a_{ij} y_j$$

and let E be the maximal error, $E = \max |E_i|$. We want E to be small.

Of course, we may immediately reduce to the case $x_j \in [0, 1]$. The methods already described give a fast algorithm for finding a simultaneous round-off with $E \le [2m \ln (2m)]^{1/2}$.

The Three Permutation Conjecture. I believe it was Jozsef Beck who first came up with the following beautiful conjecture.

Three Permutation Conjecture. Let $\sigma_1, \sigma_2, \sigma_3$ be permutations on $[n]$. Then there exists a two-coloring $\chi: [n] \to \{-1, +1\}$ such that

$$\left| \sum_{j=1}^{t} \chi(\sigma_i(j)) \right| \le K$$

for all $i = 1, 2, 3$ and all t, $1 \le t \le n$.

Here K must be an absolute constant, independent of n.

Let $\mathscr{A}_i$ be the family of intervals $\{\sigma_i(1), \cdots, \sigma_i(t)\}$. The Three Permutation Conjecture then says disc $(\mathscr{A}_1 \cup \mathscr{A}_2 \cup \mathscr{A}_3) \le K$. In fact, herdisc $(\mathscr{A}_1 \cup \mathscr{A}_2) \le 1$. Say for convenience that n is even and fix σ_1, σ_2, for example,

$$\begin{array}{lllll} \sigma_1 & 1\text{–}2 & 3\text{–}4 & 5\text{–}6 & 7\text{–}8 \\ \sigma_2 & 3\text{–}8 & 2\text{–}6 & 1\text{–}5 & 4\text{–}7. \end{array}$$

(We can assume σ_1 is the identity without loss of generality.) Consider the graph G on vertices $[n]$ given by joining $\sigma_i(2j-1)$ to $\sigma_i(2j)$ for $i = 1, 2$ and j even. Each point has degree two so G consists of cycles. The edges of the cycles

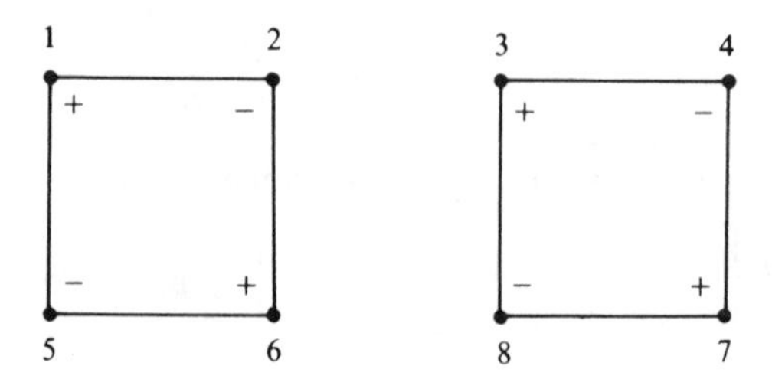

alternate between the two permutations, so they must be even. Now color each cycle alternately + and −:

$$\begin{array}{cccccccc} + & - & + & - & - & + & + & - \\ 1 & 2 & 3 & 4 & 5 & 6 & 7 & 8 \\ 3 & 8 & 2 & 6 & 1 & 5 & 4 & 7 \\ + & - & - & + & + & - & - & + \end{array}$$

Each permutation breaks into $+-$ or $-+$ pairs and no partial sum is more than one.

The Three Permutation Conjecture is stunning in its simplicity of statement, yet total resistance to all known methods of attack. Following a well-known precedent I offer $100.00 for the resolution of this conjecture.

Convex sets. Let $\mathscr{A} \subseteq 2^{\Omega}$ with $\Omega = [n]$. Another approach to discrepancy is given by letting

$$U = U_{\mathscr{A}} = \left\{ x \in R^n : \left| \sum_{a \in A} x_a \right| \leq 1 \text{ for all } A \in \mathscr{A} \right\}.$$

Then U is a convex centrally symmetric body and the notions of discrepancy can be described in terms of U without reference to $\mathscr{A}$. Place copies of U centered at each vertex $x \in \{0, 1\}^n$ of the n-cube. Blow up the copies by t. Then

$\operatorname{disc}(U) = \min t$: $(\frac{1}{2}, \cdots, \frac{1}{2})$ is covered by the tU,

$\operatorname{lindisc}(U) = \min t$: $[0, 1]^n$ is covered by the tU,

$\operatorname{herdisc}(U) = \min t$: $\{0, \frac{1}{2}, 1\}^n$ is facecovered by the tU,

$\operatorname{herlindisc}(U) = \min t$: $[0, 1]^n$ is facecovered by the tU.

Here we say $x = (x_1, \cdots, x_n)$ is facecovered by the tU if it lies in some $tU + y$, where $y = (y_1, \cdots, y_n) \in \{0, 1\}^n$ satisfies $y_i = x_i$ whenever $x_i = 0$ or 1. That is, x

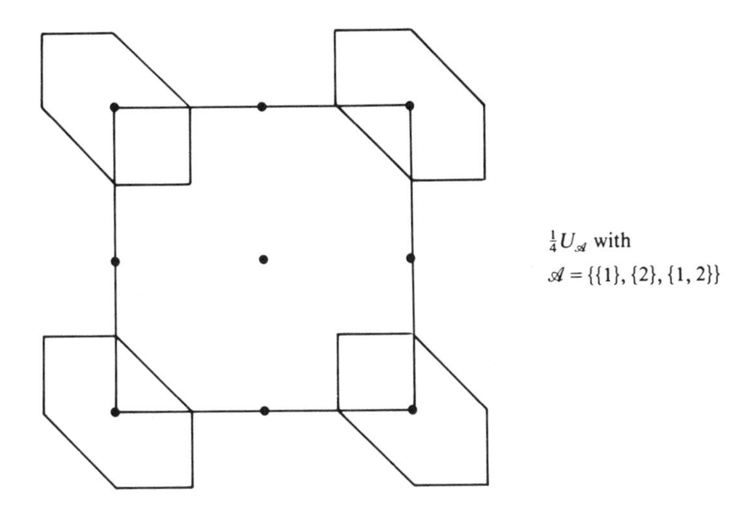

$\frac{1}{4}U_{\mathscr{A}}$ with $\mathscr{A} = \{\{1\}, \{2\}, \{1, 2\}\}$

must be covered from a copy of tU centered at some vertex y of the minimal face of x.

REFERENCE

Much of this material is in:

L. Lovász, J. Spencer and K. Vesztergombi, *Discrepancy of set-systems and matrices*, European J. Combin., 7 (1986), pp. 151–160.

LECTURE 6

Chaos from Order

Gale-Berlekamp switching game. In the corner of a third-floor office at AT&T Bell Laboratories lies a game designed and built by Elwyn Berlekamp a quarter century ago. The game consists of an 8×8 array of lights and sixteen switches, one for

$$
\begin{array}{ccccccccc}
 & \lceil & \lceil & \lceil & \lceil & \lceil & \lceil & \lceil & \lceil \\
\lceil & + & + & + & - & + & + & + & - \\
\lceil & - & - & + & - & - & + & + & + \\
\lceil & - & - & + & - & - & + & - & - \\
\lceil & - & + & + & - & + & + & - & - \\
\lceil & - & + & + & - & + & - & + & + \\
\lceil & + & + & - & - & + & - & + & + \\
\lceil & - & + & - & + & + & + & - & + \\
\lceil & + & + & + & + & + & - & + & +
\end{array}
\qquad \lceil = \text{switches}
$$

each row and column. Each switch when thrown changes each light in its line from off to on or from on to off. The object is to turn on as many lights as possible. (There are 64 switches in the back with which any initial position can be set, but to use them is cheating!) David Gale also invented this game, which is called the Gale–Berlekamp switching game. In the example above, with $+$ meaning on and $-$ meaning off, the best result is achieved by switching the third and sixth columns and the second, third and fourth rows, leaving 48 lights on.

Let the initial configuration be given by a matrix $A = [a_{ij}]$, where $a_{ij} = +1$ if the light in position (i, j) is on; otherwise $a_{ij} = -1$. Some reflection shows that the order in which the switches are pulled is immaterial; it only matters which are pulled. Set $x_i = -1$ if the ith row switch is pulled, otherwise $x_i = +1$. Set $y_j = -1$ if the jth column switch is pulled, otherwise $y_j = +1$. After we pull switches the light in position (i, j) has value $a_{ij}x_iy_j$, and so $\sum_{i=1}^{8}\sum_{j=1}^{8} a_{ij}x_iy_j$ gives #(Lights On) − #(Lights Off). We want to maximize this quantity.

We shall generalize to square (for convenience) matrices of size n and set

$$F(n)=\min_{a_{ij}}\max_{x_i,y_j}\sum_{i=1}^{n}\sum_{j=1}^{n}a_{ij}x_iy_j$$

where i, j range from 1 to n and a_{ij}, x_i, y_j range over ±1. The methods of Lecture 1 rapidly give an upper bound on $F(n)$.

THEOREM. *$F(n)\le cn^{3/2}$ with $c=2(\ln 2)^{1/2}$.*

Proof. Let $a_{ij}=\pm1$ be randomly chosen. For any choice of x_i, y_j the terms $a_{ij}x_iy_j$ are mutually independent, each equiprobably ±1 so that $\sum a_{ij}x_iy_j$ has distribution S_{n^2}. When we examine the extreme tail

$$\Pr\left[\sum a_{ij}x_iy_j>\lambda\right]<e^{-\lambda^2/2n^2}=2^{-2n}$$

with $\lambda=cn^{3/2}$. There are only 2^{2n} choices of x_i, y_j so

$$\Pr\left[\sum a_{ij}x_iy_j>\lambda \text{ for some } x_i, y_j\right]<2^{2n}2^{-2n}=1.$$

Hence there is a point in the probability space, a specific matrix a_{ij}, such that $\sum a_{ij}x_iy_j\le\lambda$ for all x_i, y_j. □

Our concern in this Lecture will be with the lower bound. Given a_{ij} we want to find (or prove the existence of) x_i, y_j that "imbalance" the matrix. The a_{ij} are arbitrary. This is not the same as saying that they are random; indeed, we may think of the a_{ij} as being selected in a pathological way by an opponent. Yet we will use probabilistic methods to scramble an arbitrary a_{ij} so that we may treat it much like a random one.

Scrambling the rows. Let a_{ij} be fixed. Let $y_j=\pm1$ be selected at random and let

$$R_i=\sum_{j=1}^{n}a_{ij}y_j$$

be the ith row sum after the switches y_j. Fixing i note that the row $(a_{i1}y_1,\cdots,a_{in}y_n)$ can be any of 2^n possibilities, all equally likely. The distribution on the row is uniform regardless of the row's initial value. Hence R_i has distribution S_n. Thus

$$E[|R_i|]=E[|S_n|]\sim cn^{1/2}$$

where $c=\sqrt{2/\pi}$. (We may approximate S_n as $n^{1/2}Y$ with Y standard normal. A closed form expression for $E[|S_n|]$ was requested in the 1974 Putnam Competition (Problem A-4). The answer is

$$E[|S_n|]=n2^{1-n}\binom{n-1}{[(n-1)/2]}$$

which we leave as a (not so easy!) exercise.)

By linearity of expectation

$$E\left[\sum_{i=1}^{n}|R_i|\right]=\sum_{i=1}^{n}E[|R_i|]\sim cn^{3/2}.$$

Note that the correlations between the R_i do depend on the original matrix.

However, linearity of expectation ignores this interplay. Hence there is a point in the probability space—a choice of $y_1, \cdots, y_n = \pm 1$—with $\sum |R_i| \geq cn^{3/2}$. Finally, pick $x_i = \pm 1$ so that $x_i R_i = |R_i|$. Then

$$\sum_{i,j} a_{ij} x_i y_j = \sum_i x_i R_i = \sum_i |R_i| \geq cn^{3/2}. \qquad \square$$

A dynamic algorithm. The methods of Lecture 4 allow us to transform the above result into a rapid algorithm. We select $y_1, \cdots, y_n$ sequentially. Having chosen $y_1, \cdots, y_{j-1}$ we choose $y_j = \pm 1$ so as to maximize

$$E\left[\sum_{i=1}^{n} |R_i| \,|\, y_1, \cdots, y_{j-1}, y_j\right].$$

To calculate this conditional expectation (recall that such calculation was the "flaw" in changing probabilistic theorems into dynamic algorithms) we set $r_i = a_{i1}y_1 + \cdots + a_{ij}y_j$ (the partial row sum to date) and note that R_i has conditional distribution $r_i + S_{n-j}$. Then

$$\begin{aligned} E\left[\sum_{i=1}^{n} |R_i| \,|\, y_1, \cdots, y_j\right] &= \sum_{i=1}^{n} E[|R_i| \,|\, y_1, \cdots, y_j] \\ &= \sum_{i=1}^{n} E[|r_i + S_{n-j}|] \\ &= \sum_{i=1}^{n} \sum_{t=0}^{n-j} \binom{n-j}{t} 2^{-(n-j)} |r_i + 2t - (n-j)| \end{aligned}$$

may be easily calculated. (Observe, however, that we have made heavy use of the restriction $a_{ij} = \pm 1$.) With this algorithm

$$\begin{aligned} &E[\sum |R_i| \,|\, y_1, \cdots, y_j] \\ &\quad = \max\left[E[\sum |R_i| \,|\, y_1, \cdots, y_{j-1}, y_j = +1], E[\sum |R_i| \,|\, y_1, \cdots, y_{j-1}, y_j = -1]\right] \\ &\quad \geq \tfrac{1}{2}\left[E[\sum |R_i| \,|\, y_1, \cdots, y_{j-1}, y_j = +1] + E[\sum |R_i| \,|\, y_1, \cdots, y_{j-1}, y_j = -1]\right] \\ &\quad = E[\sum |R_i| \,|\, y_1, \cdots, y_{j-1}]. \end{aligned}$$

As this holds for all j,

$$E[\sum |R_i| \,|\, y_1, \cdots, y_n] \geq E[\sum |R_i|] \sim cn^{3/2}.$$

With $y_1, \cdots, y_n$ fixed, the conditional expectation is simply the value. That is, we have found specific $y_1, \cdots, y_n$ for which

$$\sum |R_i| \geq E[\sum |R_i|] \sim cn^{3/2}.$$

Fixing x_i with $x_i R_i = |R_i|$, we complete the algorithm.

A parallel algorithm. Now we outline an efficient parallel algorithm to find x_i, y_j with

$$\sum_{i=1}^{n} \sum_{j=1}^{n} a_{ij} x_i y_j \geq n^{3/2}/\sqrt{3}.$$

Note this is somewhat weaker than the $n^{3/2}\sqrt{2/\pi}$ guaranteed above.

LEMMA. Let $Y_1, \cdots, Y_m = \pm 1$ be 4-wise independent random variables, each ± 1 with probability .5. Set $R = Y_1 + \cdots + Y_n$. Then

$$E[|R|] \geq \sqrt{\frac{n}{3}}.$$

Here is an argument due to Bonnie Berger. We calculate

$$E[R^2] = \sum E[Y_i Y_j] = \sum_i E[Y_i^2] = n.$$

$$E[R^4] = \sum E[Y_i Y_j Y_k Y_l] = \sum_i E[Y_i^4] + \sum_{i<j} 6E[Y_i^2 Y_j^2] = 3n^2 - 2n \leq 3n^2.$$

We're using here, for example, that for distinct i, j, k, l the expectation of the product $Y_i Y_j Y_k Y_l$ is (by 4-wise independence) the product of the expectations, which is zero. For all x "simple calculus" yields

$$|x| \geq \frac{\sqrt{3}}{2\sqrt{n}}\left(x^2 - \frac{x^4}{9n}\right),$$

though we admit the motivation is not transparent here! Then

$$E[|R|] \geq \frac{\sqrt{3}}{2\sqrt{n}}\left[E[R^2] - \frac{E[R^4]}{9n}\right] \geq \frac{\sqrt{3}}{2\sqrt{n}}\left[n - \frac{3n^2}{9n}\right] = \sqrt{n/3}$$

as desired. While the precise constant is difficult to motivate let us note that the condition of 4-wise independence is "close to" complete independence and so one should expect R to be "close to" the sum of n independent Y_i so that $E[|R|]$ should be close to $\sqrt{2n/\pi}$.

Note that the same result holds for $R_i = a_{i1}Y_1 + \cdots + a_{in}Y_n$ with all $a_{ij} = \pm 1$ since if $a_{ij} = -1$ one could simply replace Y_i by $-Y_i$. Thus if $y_1, \cdots, y_n = \pm 1$ are chosen from any 4-wise independent distribution and then x_i is chosen with $x_i R_i = |R_i|$ we will have, as previously,

$$E[R] = nE[|R_1|] \geq \frac{n^{3/2}}{\sqrt{3}}.$$

We use a fact about small sample spaces. There is a set $\Omega \subset \{-1, +1\}^n$ of size $O(n^4)$ with the following property: Let Y_i be the ith coordinate of a uniformly chosen element of Ω. Then all $E[Y_i] = 0$ and the Y_i are 4-wise independent.

Now we can describe the parallel algorithm. We have $O(n^4)$ banks of processors, each bank corresponding to a particular $(y_1, \cdots, y_n) \in \Omega$. For each such vector value we compute the R_i, the x_i and the final R. We pick that vector (and its corresponding x_i) which gives the largest R. As the average value of R over the sample space is at least $n^{3/2}/\sqrt{3}$, the R that we get will be at least that large.

Edge discrepancy. As in Lecture 1 let $g(n)$ be the minimal integer so that if K_n is Red-Blue colored there exists a set of vertices S on which the number of Red edges differs from the number of Blue edges by at least $g(n)$. In Lecture 1 we showed $g(n) \le cn^{3/2}$ by elementary probabilistic means. Now we give a lower bound essentially as a corollary of the Gale–Berlekamp switching game result.

THEOREM. $g(n) > c'n^{3/2}$.

For convenience consider $2n$ points $T \cup B$ with $T = \{T_1, \cdots, T_n\}$, $B = \{B_1, \cdots, B_n\}$ and fix an edge coloring χ with values ± 1. On the bipartite $T \times B$ we associate an $n \times n$ matrix $a_{ij} = \chi(T_i, B_j)$. Fix $x_i, y_j = \pm 1$ with $\sum a_{ij}x_iy_j > cn^{3/2}$. Let T^+, T^- denote those T_i with corresponding $x_i = +1, -1$, respectively, and B^+, B^- those B_j with corresponding $y_j = +1, -1$, respectively. Then

$$cn^{3/2} < \sum a_{ij}x_iy_j = \chi(T^+, B^+) - \chi(T^+, B^-) - \chi(T^-, B^+) + \chi(T^-, B^-).$$

(Here $\chi(X, Y) = \sum \chi(x, y)$, the summation over $x \in X$, $y \in Y$.) One of the four addends must be large, with $T^* = T^+$ or T^- and $B^* = B^+$ or B^-:

$$|\chi(T^*, B^*)| > cn^{3/2}/4.$$

To transfer this bipartite discrepancy into a complete graph discrepancy we use the equality

$$\chi(T^* \cup B^*) = \chi(T^*) + \chi(B^*) + \chi(T^*, B^*).$$

(Here $\chi(X) = \sum \chi(x_1, x_2)$, the summation over all $\{x_1, x_2\} \subset X$.) For $S = T^*, B^*$, or $T^* \cup B^*$

$$g(2n) \ge |\chi(S)| > cn^{3/2}/12. \qquad \square$$

Moreover, given χ our methods give a rapid algorithm to find an S with this discrepancy.

Tournament ranking. Using the notation of Lectures 1 and 2 we now show that $F(n) > cn^{3/2}$. That is, for any tournament T on n players there exists a ranking σ for which

$$\#[\text{nonupsets}] - \#[\text{upsets}] > cn^{3/2}.$$

For convenience assume there are $2n$ players $T \cup B$, $T = \{T_1, \cdots, T_n\}$, $B = \{B_1, \cdots, B_n\}$ and define an $n \times n$ matrix by

$$a_{ij} = \begin{cases} +1 & \text{if } T_i \text{ beats } B_j, \\ -1 & \text{if } B_j \text{ beats } T_i. \end{cases}$$

As before, we find x_i, $y_j = \pm 1$ with $\sum a_{ij}x_iy_j > cn^{3/2}$. We define T^+, T^-, B^+, B^- so that for some T^*, B^*

$$\left| \sum_{i \in T^*} \sum_{j \in B^*} a_{ij} \right| > cn^{3/2}.$$

Suppose the summation positive (otherwise reverse T and B). Let $R=(T\cup B)-(T^*\cup B^*)$, the "rest" of the players. Internally rank T^*, B^*, R so that #[nonupsets] ≥ #[upsets] on each set. Either rank $R<T^*\cup B^*$ or $R>T^*\cup B^*$ so that nonupsets outnumber upsets on games between R and $T^*\cup B^*$. Rank $T^*<B^*$. On games between T^* and B^*

$$\#[\text{nonupsets}]-\#[\text{upsets}]=\sum_{i\in T^*}\sum_{j\in B^*}a_{ij}>cn^{3/2}.$$

Summing over all games, we have

$$\#[\text{nonupsets}]-\#[\text{upsets}]\geq cn^{3/2}+0+0+0+0=cn^{3/2}.$$

Again observe that we have a polynomial time algorithm for finding this "good" ranking.

REFERENCES

The Gale-Berlekamp switching game, tournament ranking, and edge discrepancy (generalized to k-graphs) are given, respectively, in:

T. BROWN AND J. SPENCER, *Minimization of ±1 matrices under line shifts*, Colloq. Math. (Poland), 23 (1971), pp. 165-171.

J. SPENCER, *Optimal ranking of tournaments*, Networks, 1 (1972), pp. 135-138.

P. ERDÖS AND J. SPENCER, *Imbalances in k-colorations*, Networks, 1 (1972), pp. 379-385.

Oddly, the connection between the switching game and the other problems was not recognized at that time. The effective algorithm is new. The tournament paper was my first "real" theorem (publication was somewhat delayed) and provided simultaneously an introduction to the probabilistic method, a first meeting with Paul Erdös, and the core of my dissertation. It is a pleasure to return to this question with a fresh, algorithmic viewpoint.

Chapter 15 of *The Probabilistic Method* (referenced in the Preface) is one of many sources on small sample spaces and parallel algorithms.

LECTURE 7

Random Graphs II

Clique number concentration. Let G be the random graph on n vertices with edge probability $p=\frac{1}{2}$. We examine the clique number $\omega(G)$. It is not difficult to show $\omega(G)\sim 2\lg n$. Indeed $\omega(G)<2\lg n$ was essentially shown in finding the basic lower bound on the Ramsey $R(k,k)$ in Lecture 1. The lower bound involves a second moment method. The result is surprisingly tight.

THEOREM. *There is a* $k=k(n)$ *so that*

$$\lim_n \Pr[\omega(G)=k \text{ or } k+1]=1.$$

In addition, $k(n)\sim 2\lg n$.

Proof. Given n let $X^{(k)}$ be the number of cliques of G of size k and set $f(k)=E[X^{(k)}]$ so that

$$f(k)=\binom{n}{k}2^{-\binom{k}{2}}.$$

Elementary calculation shows that $f(k)$ increases until $k\sim\lg n$ and then decreases. Let k_0 be the first value where $f(k)$ goes under one, i.e.,

$$f(k_0-1)\geq 1>f(k_0).$$

Then $k_0\sim 2\lg n$ (Lecture 1) and f is quite steep in this range. For $k\sim 2\lg n$

$$f(k+1)/f(k)\sim(n/k)2^{-k}<1/n.$$

Basically we show that if $f(k)\ll 1$, then G will not have a clique of size k, whereas if $f(k)\gg 1$, then G will have a clique of size k. One part is easy. As the "logarithmic derivative" is so steep

$$f(k_0+1)<f(k_0)/n\ll 1,$$

so

$$\begin{aligned}\Pr[\omega(G)\geq k_0+1]&=\Pr[X^{(k_0+1)}\neq 0]\\&\leq E[X^{(k_0+1)}]=f(k_0+1)\ll 1.\end{aligned}$$

Now let $k=k_0-2$ so that (with $X=X^{(k)}$ for convenience)

$$E[X]=f(k)>f(k+1)n \gg 1.$$

We apply the second moment method and the notation of Lecture 3. Write $X=\sum X_S$, where the summation ranges over all k-element subsets of $[n]$ and X_S is the indicator random variable for the event "S is a clique." Let

$$m=\binom{n}{k}=\text{number of } S, \qquad \mu=2^{-\binom{k}{2}}=E[X_S].$$

If $|S\cap T|=i$, then

$$E[X_T|X_S=1]=E[X_T]2^{\binom{i}{2}}.$$

(Knowing $X_S=1$ puts $\binom{i}{2}$ "free" edges into T.)

$$f(S,T)=\frac{E[X_T|X_S=1]}{E[X_T]}-1=2^{\binom{i}{2}}-1.$$

Fix S and let $f(T)=f(S,T)$. We must show

$$E_T[f(T)]=o(1). \tag{??}$$

Splitting T according to $i=|S\cap T|$, we have that

$$E_T[f(T)]=\sum_{i=0}^{k} g(i)$$

where

$$g(i)=\Pr[|S\cap T|=i](2^{\binom{i}{2}}-1).$$

For $i=0$ or 1, $g(i)=0$, there is no correlation. For other i we ignore the "-1" term. When $i=k$, $g(k)<1/E(X)=o(1)$. (Recall this always happens.) For $i\le k/2$ we roughly bound

$$g(i)<k^i(k/n)^i 2^{i^2/2}=[k^2 2^{i/2}/n]^i \ll 1.$$

(This bound is actually good up to $i<k(1-\varepsilon)$.) For $j\le k/2$ calculation gives

$$\begin{aligned} g(k-j) &< k^j n^j 2^{\binom{k-j}{2}-\binom{k}{2}} g(k) \\ &< (kn)^j 2^{-jk} 2^{j(j+1)/2} \\ &= [kn2^{(j+1)/2}2^{-k}]^j \ll 1. \end{aligned}$$

Each $g(i)$ is small; with a moderate amount of technical work its sum is small, the second moment method applies and

$$\Pr[\omega(G)<k]=\Pr[X=0]\ll 1.$$

Almost always $\omega(G)$ is either k_0-2, k_0-1 or k_0.

Let us eliminate one of the possibilities. As $f(k_0)/f(k_0-1)\ll 1$ either $f(k_0)\ll 1$ or $f(k_0-1)\gg 1$. In the first case almost always $\omega(G)<k_0$ so $\omega(G)=k_0-2$ or k_0-1. In the second case almost always $\omega(G)\ge k_0-1$ so $\omega(G)=k_0-1$ or k_0.

Actually, for "most" n the function $f(k)$ jumps over one in that $f(k_0)\ll 1$ *and* $f(k_0-1)\gg 1$. For these n the clique number $\omega(G)$ is almost certainly k_0-1.

Chromatic number. Again let G have n vertices with $p=\frac{1}{2}$. Then $\bar{G}$ has the same distribution, so almost always $\omega(\bar{G})\sim 2\lg n$ and

$$\chi(G)\geq n/\omega(\bar{G})\sim n/(2\lg n).$$

How about an upper bound—a coloring of G? Here is a simple algorithm: Select sequentially points $P_1, P_2, \cdots$ with $P_1=1$ and P_i the smallest point independent of $P_1, \cdots, P_{i-1}$. It is not difficult to show that an independent set of size roughly $\lg n$ is selected in this way. (Roughly each point selected eliminates half of the pool of available points.) Color these points "1," throw them away and iterate.

Central to the analysis of this algorithm is that the remaining graph G' has the distribution of a random graph. To see this, imagine that the initial graph is "hidden." We pick $P_1=1$ and then "expose" all edges P_1Q. Then we pick P_2 and expose P_2Q, etc. When the procedure terminates with $P_1, P_2, \cdots$ found, none of the edges in G' have yet been exposed; hence they may still be considered random.

When the procedure is iterated an independent set of size $\lg m$ is found when m points remain. Thus $n/(\lg n)$ colors will be needed, asymptotically, for the entire graph. That is,

$$n/(2\lg n)<\chi(G)<n/(\lg n).$$

Since the Durango Lectures Béla Bollobás has shown that the lower bound is asymptotically correct. A proof is given in the Bonus Lecture.

Note that the following does *not* work: Take an independent set of size $2\lg n$, color it "1," delete it, and iterate. The problem is that the remaining G' *cannot* be treated as a random graph.

Sparse graphs. Let G have n vertices with edge probability $p=n^{\varepsilon-1}$. Let $X^{(k)}$ be the number of k-cliques in $\bar{G}$. Then

$$E[X^{(k)}]=\binom{n}{k}(1-p)^{\binom{k}{2}}$$

$$\sim\left(\frac{ne}{k}\right)^k e^{-pk^2/2}=\left[\frac{ne}{k}e^{-pk/2}\right]^k.$$

For $k>(2\varepsilon\ln n)/p$, $E[X^{(k)}]\ll 1$ and so $\omega(\bar{G})\leqq k$. Set $d=np=n^{\varepsilon}$, the average degree. Then

$$\chi(G)>n/k\sim d/(2\ln d).$$

The selection procedure described for $p=\frac{1}{2}$ works also in this case and gives

$$\chi(G)<d/(\ln d).$$

An early gem. I find the probabilistic method most striking when the statement of the theorem does not appear to call for its use. The following result of Paul Erdös is a favorite of mine.

THEOREM. *For all K, L there exists a graph G with*

$$\chi(G) > L, \qquad \text{girth}\,(G) > K.$$

Proof. Fix positive $\varepsilon < 1/k$ and let G be the random graph with n vertices and edge probability $p = n^{\varepsilon - 1}$. We have just shown that

$$(*) \qquad \omega(\bar{G}) < cn^{1-\varepsilon}(\ln n)$$

almost always. Let Y be the number of cycles of length at most k in G. For $3 \le i \le k$ there are $(n)_i/2i$ potential i-cycles each in G with probability p^i so

$$E[Y] = \sum_{i=3}^{k} [(n)_i/2i]p^i = \sum_{i=3}^{k} n^{\varepsilon i}/2i = o(n).$$

Hence

$$(**) \qquad Y < n/2$$

almost always. Pick n sufficiently large so that $(*)$ and $(**)$ each occur with probability greater than $\frac{1}{2}$. With positive probability both hold. That is, there exists a specific graph G on n vertices satisfying $(*)$ and $(**)$. Delete from G one vertex of each cycle of size at most k, leaving a subgraph G'. Clearly girth $(G') > k$ as we have destroyed the small cycles. At most Y points were deleted so G' has at least $n/2$ vertices. Also $\omega(\bar{G}') \le \omega(\bar{G}) = cn^{1-\varepsilon}(\ln n)$ so that

$$\chi(G') > (n/2)/(cn^{1-\varepsilon} \ln n) = kn^{\varepsilon}/\ln n.$$

Pick n sufficiently large so that this lower bound is at least L. $\square$

Concentration via martingales. The martingale method allows one to prove that a graph function is tightly concentrated in distribution, though it does not say where this concentration occurs. A martingale is a stochastic process $X_0, \cdots, X_n$ for which $E[X_{i+1}|X_i] = X_i$. The following bound on deviations of martingales follows the argument of Lecture 4.

THEOREM. *Let* $0 = X_0, X_1, \cdots, X_n$ *be a martingale with* $|X_{i+1} - X_i| \le 1$. *Then*

$$\Pr[X_n > \lambda] < e^{-\lambda^2/2n}.$$

Proof. Set $Y_i = X_i - X_{i-1}$. If Y is any distribution with $E(Y) = 0$ and $|Y| \le 1$ the concavity of $f(y) = e^{\alpha y}$ implies

$$E[e^{\alpha Y}] \le \cosh(\alpha) \le e^{\alpha^2/2}.$$

In particular,

$$E[e^{\alpha Y_i}|Y_1, \cdots, Y_{i-1}] < e^{\alpha^2/2}.$$

Hence

$$E[e^{\alpha X_n}] = E\left[\prod_{i=1}^{n} e^{\alpha Y_i}\right] < e^{\alpha^2 n/2}$$

and with $\alpha = \lambda/n$

$$\Pr[X_n > \lambda] = E[e^{\alpha X_n}]\, e^{-\alpha\lambda} < e^{\alpha^2 n - \alpha\lambda} = e^{-\lambda^2/2n}. \qquad \square$$

For those unfamiliar with martingales the gambling analogy may be helpful. Imagine a fair casino in which there are a variety of games available—flipping a coin, roulette (without 0 or 00), not playing—each with zero expectation. A player enters with X_0 dollars and plays n rounds. His choice of game may depend on previous success: "If I lose three times at roulette I'll switch to flipping a coin" or "If I'm ahead \$50.00 I'll quit." Assume that the maximum gain or loss on a single round is \$1.00. The above theorem bounds a player's chance of ending up λ dollars ahead. The precise result, with λ integral, is that the player's optimal strategy is to play "flip a coin" until he is ahead λ dollars and then quit.

Let G be the random graph with vertex set $[n]$ and edge probability $p = \frac{1}{2}$. For $1 \le i \le n$ define a function X_i with domain the set of graphs on $[n]$ by

$$X_i(H) = E[\chi(G) \mid \quad G|_{[i]} = H|_{[i]}].$$

That is, $X_i(H)$ is the predicted value of $\chi(H)$ when we know H on $[i]$ and all other edges are in H with independent probability $\frac{1}{2}$. At the extremes $X_1(H) = E[\chi(G)]$ is a constant independent of H and $X_n(H) = \chi(H)$. The sequence $X_1, \cdots, X_n$ moves toward the actual chromatic number as more and more of H is "uncovered." Considering H in the probability space of random graphs the sequence $X_1, \cdots, X_n$ forms a martingale, part of a general class known as a Doob Martingale Process. Basically, given that with $[i]$ uncovered the predicted value of $\chi(H)$ is X_i, then the average value of the value predicted when $i+1$ is uncovered is still X_i. A somewhat similar situation occurred in Lecture 6, where $E[\sum |R_i| \mid y_1, \cdots, y_{j-1}]$ is the average of the possible $E[\sum |R_i| \mid y_1, \cdots, y_j]$. Also $|X_{i+1} - X_i| \le 1$. For any H knowing about the $(i+1)$st point can only change $E[\chi(H)]$ by one as, essentially, one point can only affect the chromatic number by at most one.

We normalize by subtracting $c = E[\chi(G)]$ and apply the bound on deviations of martingales.

THEOREM. *Let H be a random graph on n vertices with edge probability $p = \frac{1}{2}$. With c as above*

$$\Pr[|\chi(H) - c| > \lambda\sqrt{n-1}] < 2\, e^{-\lambda^2/2}.$$

With $\lambda = \omega(n) \to \infty$ arbitrarily slow this means that the distribution of chromatic number is concentrated in an interval of width $\sqrt{n}\,\omega(n)$. Note, however, that we cannot say where this interval is, so that the bounds on $\chi(G)$ previously given remain open.

This proof applies for any value of p. If $p = p(n)$ approaches zero a refinement of this method gives an even tighter concentration. The strongest result occurs when $p = n^{-a}$ and $a > 5/6$. Then $\chi(G)$ is concentrated on five values—there is a $k = k(n)$ (but we do not know what it is!) so that almost always $\chi(G)$ is either $k, k+1, k+2, k+3$ or $k+4$.

REFERENCES

The Bollobás and Palmer books cited in Lecture 3 again provide basic references. Large girth and large chromatic number are in:

P. ERDÖS, *Graph theory and probability*, Canad. J. Math., 11 (1959), pp. 34-38.

For the martingale results, see:

E. SHAMIR AND J. SPENCER, *Sharp concentration of the chromatic number on random graphs* $G_{n,p}$, Combinatorica, 7 (1987).

LECTURE 8

The Lovász Local Lemma

The Lemma. Let $A_1, \cdots, A_n$ be events in a probability space. In combinatorial applications the A_i are "bad" events. We wish to show $\Pr[\bigwedge \bar{A}_i] > 0$ so that there is a point (coloring, tournament, configuration) x which is good. The basic probabilistic method of Lecture 1 may be written:

Counting sieve. If $\sum \Pr[A_i] < 1$, then $\Pr[\bigwedge \bar{A}_i] > 0$.

There are other simple conditions that ensure $\Pr[\bigwedge \bar{A}_i] > 0$.

Independence sieve. If $A_1, \cdots, A_n$ are mutually independent and all $\Pr[A_i] < 1$, then $\Pr[\bigwedge \bar{A}_i] > 0$.

The Lovász Local Lemma is a sieve method which allows for *some* dependence among the A_i. A graph G on vertices $[n]$ (the indices for the A_i) is called a *dependency graph* for $A_1, \cdots, A_n$ if for all i A_i is mutually independent of all A_j with $\{i, j\} \notin G$. (That is, A_i is independent of any Boolean function of these A_j.)

LOVÁSZ LOCAL LEMMA (Symmetric case). *Let $A_1, \cdots, A_n$ be events with dependency graph G such that*

$$\Pr[A_i] \le p \quad \textit{for all } i, \qquad \deg(i) \le d \quad \textit{for all } i$$

and

$$4dp < 1.$$

Then

$$\Pr[\bigwedge \bar{A}_i] > 0.$$

Proof. We show by induction on s that if $|S| \le s$, then for any i

$$\Pr\left[A_i \,\middle|\, \bigwedge_{j \in S} \bar{A}_j\right] \le 2p.$$

For $S = \emptyset$ this is immediate. Renumber for convenience so that $i = n$, $S = \{1, \cdots, s\}$ and $\{i, x\} \notin G$ for $x > d$. Now

$$\Pr[A_n | \bar{A}_1 \cdots \bar{A}_s] = \frac{\Pr[A_n \bar{A}_1 \cdots \bar{A}_d | \bar{A}_{d+1} \cdots \bar{A}_s]}{\Pr[\bar{A}_1 \cdots \bar{A}_d | \bar{A}_{d+1} \cdots \bar{A}_s]}.$$

We bound the numerator

$$\Pr[A_n\bar{A}_1 \cdots \bar{A}_d|\bar{A}_{d+1} \cdots \bar{A}_s] \le \Pr[A_n|\bar{A}_{d+1} \cdots \bar{A}_s]$$
$$= \Pr[A_n] \le p$$

as A_n is mutually independent of $A_{d+1}, \cdots, A_s$. We bound the denominator

$$\Pr[\bar{A}_1 \cdots \bar{A}_d|\bar{A}_{d+1} \cdots \bar{A}_s] \ge 1 - \sum_{i=1}^{d} \Pr[A_i|\bar{A}_{d+1} \cdots \bar{A}_s]$$
$$\ge 1 - \sum_{i=1}^{d} 2p \qquad \text{(Induction)}$$
$$= 1 - 2pd \ge \tfrac{1}{2}.$$

Hence we have the quotient

$$\Pr[A_n|\bar{A}_1 \cdots \bar{A}_s] \le p/\tfrac{1}{2} = 2p,$$

completing the induction. Finally

$$\Pr[\bar{A}_1 \cdots \bar{A}_n] = \prod_{i=1}^{n} \Pr[\bar{A}_i|\bar{A}_1 \cdots \bar{A}_{i-1}] \ge \prod_{i=1}^{n} (1-2p) > 0. \qquad \square$$

The proof is so elementary that it could, and I think should, be taught in a first course in probability. It has had and continues to have a profound effect on probabilistic methods.

The diagonal Ramsey function. A lower bound for $R(k, k)$, our first use of the probabilistic method in Lecture 1, provides a simple application of the Lovász Local Lemma. Consider a random two-coloring of K_n with A_S the event "S is monochromatic," S ranging over the k-sets of vertices. Define G by placing $\{S, T\} \in G$ if and only if $|S \cap T| \ge 2$. Then A_S is mutually independent of all A_T with T not adjacent to G, since the A_T give information only about edges outside of S. Hence G is a dependency graph. (When $|S \cap T| = 2$ the events A_S, A_T are independent; note however that mutual independence from a family of A_T is far stronger than pairwise independence with each A_T. Recall the old chestnut: I have two children and at least one is a girl. What is the probability they are both girls. Conditional on the younger being a girl it is one half. Conditional on the older being a girl it is one half. Conditional on the disjunction it is one third. But I digress.) We apply the Lovász Local Lemma with $p = \Pr[A_S] = 2^{1-\binom{k}{2}}$ and

$$d = |\{T: |S \cap T| \ge 2\}| \le \binom{k}{2}\binom{n}{k-2}.$$

COROLLARY. *If*

$$4\binom{k}{2}\binom{n}{k-2}2^{1-\binom{k}{2}} < 1,$$

then $R(k, k) > n$.

The asymptotics are somewhat disappointing.

COROLLARY.

$$R(k,k) > \frac{\sqrt{2}}{e} k 2^{k/2}(1+o(1)).$$

This improves the lower bound given in Lecture 1 by a factor of 2 and the improvement via the deletion method in Lecture 2 (which, oddly, was only published after the better bound) by a factor of $\sqrt{2}$. The gap between the upper and lower bounds has not really been effectively decreased. The lower bound of Erdös was found in April 1946 (published in 1947) and progress on this difficult problem has been slow.

The van der Waerden function. Here the improvement is more impressive. Color $[n]$ randomly. For each arithmetic progression S of size k let A_S be the event that S is monochromatic. Let S, T be adjacent in G if they intersect. (In all our applications the probability space will be a random coloring of some set Ω. For Ramsey's Theorem Ω was the edge set of K_n. Events will be not adjacent if they deal with the coloring on disjoint sets.) Now $p = 2^{1-k}$ and $d \le nk$ as one progression intersects (exercise) at most nk others. Hence we have the following theorem.

THEOREM. *If* $4nk2^{1-k} < 1$, *then* $W(k) > n$. *That is,* $W(k) > 2^k/8k$.

This greatly improves the bound $W(k) > 2^{k/2}$ of Lecture 1. Still we must in all honesty mention that $W(p) \ge p2^p$, for p prime, has been shown by completely constructive means!

Algorithm? The Lovász Local Lemma proves the existence of an x satisfying $\bigwedge \bar{A}_i$ even when $\Pr[\bigwedge \bar{A}_i]$ may be exponentially small. We have seen in Lecture 4 that when $\sum \Pr[A_i] < 1$ there often is an algorithm to find a specific "good" x.

Open Problem. Can the Lovász Local Lemma be implemented by a Good Algorithm?

Let us be more specific. Suppose $S_1, \cdots, S_n \subset [n]$ with all $|S_i| = 10$ and all $\deg(j) = 10$. Two color $[n]$ randomly and let A_i be the event that S_i is monochromatic. Let i, i' be adjacent in the dependency graph if and only if their corresponding sets intersect. Each S_i is intersected by at most 90 other $S_{i'}$. We apply the Lovász Local Lemma with $p = \Pr[A_i] = 2^{-9}$ and $d = 90$. As $4dp < 1$, $\Pr[\bigwedge \bar{A}_i] > 0$ and so there is a two-coloring χ for which no S_i is monochromatic. Is there a polynomial (in n) time algorithm for finding such a coloring?

Notice that the Lovász Local Lemma here guarantees the existence of a "needle in a haystack." If, say, $S_1, \cdots, S_{n/10}$ are disjoint a random χ is good with probability at most $(1-2^{-9})^{n/10}$. Can we actually find this exponentially small needle in polynomial time?

Algorithm? Sometimes! Since the Durango lectures Jozsef Beck has made a breakthrough on algorithmic implementation of the Lovász Local Lemma. His method doesn't always work—in particular, the above specific question remains open. Let's change it slightly: Suppose $S_1, \cdots, S_n \subset [n]$ with all $|S_i| = 200$ and all $\deg(j) = 200$. We find a Red/Blue coloring with no monochromatic S_i in expected polynomial time.

First color $[n]$ randomly Red and Blue. Call S_i *bad* if at least 180 points are the same color. Uncolor all points in all bad sets, leaving a partial coloring. If S_i now has both Red and Blue points remove it. Otherwise we say S_i survived and let S_i^* be the set of uncolored points of S_i. Let F^* be the family of the S_i^*. It will suffice to color the now uncolored points so that no set in F^* is monochromatic.

If S_i was bad then all of S_i was uncolored and $S_i^* = S_i$. For S_i to survive otherwise it must intersect bad S_k. The points of $S_i - S_i^*$ are all the same color (otherwise S_i is removed) and so there must be less than 180 of them (otherwise S_i is bad) and so S_i^* has at least 20 points. Thus all sets in F^* have at least 20 points.

Replace each S_i^* with more than 20 points by a subset S_i^{**} with exactly 20 points, giving a family F^{**}. It suffices to color the uncolored points so that no set in F^{**} is monochromatic. But each vertex is still in at most 200 sets in F^{**} and so each set intersects at most 4000 other sets in F^{**}. As $4(4000)2^{1-20} < 1$ (by our judicious selection of 180) the Lovász Local Lemma assures us that the desired coloring *exists*. The problem remains: How do we find it.

The key is that F^{**} breaks into small components, with all components of size $O(\ln n)$. Let's start with an intuitive (though invalid) argument. The probability of a set being bad is twice the probability of at least 180 heads in 200 tosses of a fair coin which is less than $3 \cdot 10^{-33}$. For a set to survive at least one of its at most 40000 neighbors must have been bad, and that holds with probability at most $2 \cdot 10^{-28}$. Make a graph with vertices the initial sets S_i, adjacency being intersection. This graph has n vertices and maximal degree at most 40000. The surviving sets form a random subgraph, each vertex surviving with probability at most $2 \cdot 10^{-28}$ so that the average set has degree less than $40000(2 \cdot 10^{-10})$ in the random subgraph. With the average degree less than one we are before the Double Jump, as discussed in Lecture 3, and all components are size $O(\ln n)$. There are several difficulties with this approach—not least of which is the dependency of survivability of S_i, S_j when they are "close together." Fortunately, these difficulties can be dealt with.

Define a graph on F^{**} in which sets are joined if they intersect. Suppose some R sets are in a single component. Each set has degree at most 4000 in the graph. Start pulling out $T_1, T_2, \cdots$ where each T_j is distance at least three from all previous T_i and distance precisely three from some previous T_i. This process only stops when all sets are within two of a previously selected set so we certainly get $T_1, \cdots, T_U$ with $U = 10^{-8}R$. Any surviving set is adjacent to some bad set so there are corresponding bad $S_1, \cdots, S_U$, S_i adjacent to T_i. As the T_i are all at least three apart the S_i are mutually nonadjacent. Consider a graph on $\{S_1, \cdots, S_U\}$ in which two sets are joined if they are within distance five in the original adjacency graph. As each T_j was within three of a previous T_i, each S_j is adjacent to a previous S_i in the new graph so that the new graph is connected and hence it contains a tree. Thus we would have: A tree T on $\{1, \cdots, U\}$ and bad sets $S_1, \cdots, S_U$, mutually nonadjacent, with S_j within five of S_i whenever j, i are adjacent in T.

There are less than 4^U trees on U points, up to isomorphism. Fix such a tree, numbered so that each $j > 1$ is adjacent to at least one $i < j$. Now count potential $(S_1, \cdots, S_U)$. There are n choices for S_1. But there are less than $(40000)^5$ choices for each S_j since it must be within five of an already chosen S_i. But the probability that any particular mutually disjoint $S_1, \cdots, S_U$ is all bad is less than $[2 \cdot 10^{-33}]^U$. So all together the expected number of such trees is less than

$$n[4 \cdot (4000)^5 \cdot 2 \cdot 10^{-33}]^U.$$

The bracketed term is less than one. (One could reduce 200 somewhat and still get this but not all the way to 10.) Thus for $U = c_1 \ln n$, c_1 an absolute constant, almost surely no such tree exists. Thus for $R = c_2 \ln n$, $c_2 = 10^8 c_1$, almost surely all components have size less than R.

Now to the main point of the algorithm. We've colored, uncolored, and then we check that all components have size at most the above $R = c_2 \ln n$. Most of the time we'll have success. Each component has at most $c_3 \ln n$ points, $c_3 = 200 c_2$, and has a coloring by the Lovász Local Lemma. Now *we try all colorings!* There are only $2^{c_3 \ln n}$, polynomially many, colorings, each takes logarithmic time to check, so in polynomial time we color each component. There are less than n components so in polynomial time all of the uncolored points are colored and no set in F^{**} is monochromatic.

A few comments. The above argument is due to Noga Alon and is somewhat different from Beck's original idea. This argument allows a parallel implementation. While we've given a probabilistic algorithm both Beck and Alon's algorithms can be completely derandomized. Beck's is actually presented in derandomized form. Finally, the above algorithm is polynomial but with a possibly quite high exponent. Suppose the 200 were replaced by, say, 2000. Now one colors, uncolors and is left with F^{**} with all component sizes $O(\ln n)$. But the parameters (for 2000 appropriately large) are such that we can again color, uncolor each component, being left with F^{****}. The component sizes of F^{****} are now $O(\ln \ln n)$. So trying all colorings takes time $2^{O(\ln \ln n)}$, polylog, for each component. Altogether, this can give a coloring algorithm whose total time is merely $O(n(\ln n)^c)$ for some constant c.

Addendum: Joke. Here we give an ironic demonstration of the power of the Lovász Local Lemma.

THEOREM. *Let S, T be finite sets with $|T| \geq 8|S|$. Then there exists a function $f: S \to T$ which is injective.*

Proof. Let f be a random function. For each $\{x, y\} \subset S$ let A_{xy} be the event $f(x) = f(y)$. Then $\Pr[A_{xy}] = 1/|T| = p$. Let $\{x, y\}$ and $\{u, v\}$ be adjacent in the dependency graph if and only if they intersect. Then the maximal degree in the dependency graph is $d = 2(|S| - 1)$. As $4dp < 1$, $\Pr[\bigwedge \bar{A}_{xy}] > 0$ and so there exists an f for which $\bar{A}_{xy}$ for all x, y. That is, f is injective.

When $|T| = 365$ and $|W| = 23$ the "birthday problem" says that f has probability less than $\frac{1}{2}$ of being injective. When $|T| = 8|S|$ the probability of a random f being injective is exponentially small. The Counting Sieve proves the existence of an injective f only when $\binom{|S|}{2} < |T|$. □

Anti van der Waerden. Here is the original use of the Lovász Local Lemma.

THEOREM. *Let k, m satisfy $4m(m-1)(1-1/k)^m < 1$. Let $S \subset R$ with $|S| = m$. Then there exists a k-coloring $\chi: R \to [k]$ so that every translate $S + t$ is k-colored. That is, for all $t \in T$ and $1 \leq i \leq k$ there exists $s \in S$ with $\chi(s + t) = i$.*

Without use of the Lovász Local Lemma it is difficult even to prove the existence of an $m = m(k)$ with this property. Notice a fundamental difference between the translation and homothety groups. Gallai's Theorem, a consequence of van der Waerden's Theorem, states that for all finite S and all finite colorings of R there is a monochromatic $S' = aS + t$.

Proof. First we let $B \subset R$ be an arbitrary finite set and k-color B so that all $S+t\subset B$ have all k colors. Color B randomly. For each t such that $S+t\subset B$ let A_t be the event that $S+t$ does not have all k colors. Then

$$\Pr[A_t] \le k(1-1/k)^m = p.$$

Let t, t' be adjacent in the dependency graph if and only if $S+t$ and $S+t'$ intersect. With given t this occurs only if $t' = +s-s'$ for distinct $s, s' \in S$, so that the dependency graph has degree at most $d = m(m-1)$. The conditions on k, m are precisely that $4dp<1$. The Lovász Local Lemma applies and $\bigwedge \bar{A}_t \neq \emptyset$; there is a k-coloring of B for which all translates of S lying in B have all k colors.

Compactness. To color all of R we need the Compactness Principle. We state this in a form convenient for us.

COMPACTNESS PRINCIPLE. *Let Ω be an infinite set, k a positive integer, and let U be a family of pairs (B, χ) where $B\subset \Omega$ is finite, $\chi: B\to[k]$ such that*

(i) *U is closed under restriction. That is, if $(B,\chi)\in U$ and $B'\subset B$ then $(B', \chi|_{B'})\in U$;*

(ii) *For all B some $(B,\chi)\in U$.*

Then there exists $\chi:\Omega\to[k]$ such that

$$(B, \chi|_B)\in U \quad \textit{for all finite } B\subset U.$$

Proof. Let X be the topological space of all $\chi:\Omega\to[k]$. Here we consider $[k]$ discrete and X has the usual product topology. That is, a basis for the open sets is given by the sets $\{\chi: \chi(b_i)=a_i, 1\le i\le s\}$ over all $s, b_1, \cdots, b_s, a_1, \cdots, a_s$. For every finite B let X_B be the set of $\chi\in X$ with $(B,\chi|_B)\in U$. By (ii) $X_B\neq\emptyset$. Splitting $\chi\in X$ according to $\chi|_B$ gives a finite ($|B|^k$) partition of X into sets both open and closed so X_B, being a finite union of such sets, is closed. For any $B_1, \cdots, B_s$ property (i) gives

$$X_{B_1}\cap\cdots\cap X_{B_s}\supset X_{B_1\cup\cdots\cup B_s}\neq\emptyset.$$

Since X is the product of compact spaces it is compact. (This is the Tikhonov Theorem, which is equivalent to the Axiom of Choice.) By the finite intersection property $\bigcap X_B\neq\emptyset$, the intersection over all finite $B\subset\Omega$. Choose $\chi\in\bigcap X_B$. For all finite B $\chi\in X_B$ so $(B,\chi|_B)\in U$ as desired. □

In applications suppose a k-coloring of Ω is desired to meet certain conditions on finite sets and it is known that any finite subset $B\subset\Omega$ may be k-colored to meet these conditions. Then the finite colorings may be joined (in a kind of inverse limit) to provide a coloring of all of Ω. In particular, as any $B\subset R$ can be k-colored, there exists a k-coloring χ of R such that all translates $S+t$ have all k colors, completing the Anti van der Waerden Theorem. □

The general case. Suppose now the A_i may have different probabilities.

LOVÁSZ LOCAL LEMMA (General case). *Let $A_1, \cdots, A_n$ be events with dependency graph G. Assume there exist $x_1, \cdots, x_n \in [0, 1)$ with*

$$\Pr[A_i] < x_i \prod_{\{i,j\} \in G} (1 - x_j)$$

for all i. Then

$$\Pr[\wedge A_i] > \prod_{i=1}^{n} (1 - x_i) > 0.$$

Proof. We show by induction on s that for all i, S with $|S| \le s$

$$\Pr\left[A_i \,\middle|\, \bigwedge_{j \in S} \bar{A}_j\right] < x_i.$$

For $s = 0$, $\Pr[A_i] < x_i \prod (1 - x_j) \le x_i$ is immediate. Now renumber so $i = n$, $S = \{1, \cdots, s\}$ and among $x \in S$, $\{i, x\} \in G$ for $x = 1, \cdots, d$.

$$\Pr[A_n | \bar{A}_1 \cdots \bar{A}_s] = \frac{\Pr[A_n \bar{A}_1 \cdots \bar{A}_d | \bar{A}_{d+1} \cdots \bar{A}_s]}{\Pr[\bar{A}_1 \cdots \bar{A}_d | \bar{A}_{d+1} \cdots \bar{A}_s]}.$$

We bound the numerator

$$\Pr[A_n \bar{A}_1 \cdots \bar{A}_d | \bar{A}_{d+1} \cdots \bar{A}_s] \le \Pr[A_n | \bar{A}_{d+1} \cdots \bar{A}_s] = \Pr[A_n]$$

as before. This time we bound the denominator a bit more carefully.

$$\Pr[\bar{A}_1 \cdots \bar{A}_d | \bar{A}_{d+1} \cdots \bar{A}_s] = \prod_{i=1}^{d} \Pr[\bar{A}_i | \bar{A}_{i+1} \cdots \bar{A}_s]$$

$$\ge \prod_{i=1}^{d} (1 - x_i) \qquad \text{(Induction)}.$$

Hence we have the quotient

$$\Pr[A_n | \bar{A}_1 \cdots \bar{A}_s] \le \Pr[A_n] \Big/ \prod_{\{n,i\} \in G} (1 - x_i) < x_i,$$

completing the induction. Finally

$$\Pr[\bar{A}_1 \cdots \bar{A}_n] = \prod_{i=1}^{n} \Pr[\bar{A}_i | \bar{A}_1 \cdots \bar{A}_{i-1}]$$

$$> \prod_{i=1}^{n} (1 - x_i) > 0. \qquad \square$$

This allows a small, generally unimportant, improvement in the symmetric case. Assume all $\Pr[A_i] \le p$ and $\deg(i) \le d$. With all $x_i = x$ the above condition becomes $\Pr[A_i] \le x(1 - x)^d$. We may optimally choose $x = 1/(d + 1)$ so that the condition is $\Pr[A_i] < d^d/(d+1)^{d+1}$. This allows us asymptotically (as $d \to \infty$) to replace 4 by e so that if $(e + o(1))dp < 1$ then $\wedge \bar{A}_i \ne \phi$. The value e has been shown to be the best possible constant for this result.

Setting $y_i = x_i/(\Pr[A_i])$ the condition for the General Case may be rewritten

$$\ln y_i > \sum_{\{i,j\} \in G} -\ln(1 - y_j \Pr[A_j]).$$

As $-\ln(1-\delta) \sim \delta$ this has roughly the meaning

$$\ln y_i > \sum_{\{i,j\} \in G} y_j \Pr[A_j].$$

In the earlier edition this replacement was incorrectly made. Now let's give it a somewhat more complicated, but correct, replacement. From the Taylor Series $-\ln(1-\delta) = \delta + \frac{1}{2}\delta^2 + 0(\delta^3)$ and calculation gives, say, $-\ln(1-\delta) < \delta(1+\delta)$ for $\delta < .1$.

COROLLARY. *Let $A_1, \cdots, A_n$ be events with dependency graph G. If there exist $y_1, \cdots, y_n$ and $\delta < .1$ with $0 < y_i < \delta/\Pr[A_i]$ satisfying*

$$\ln y_i > (1+\delta) \sum_{\{i,j\} \in G} y_j \Pr[A_j]$$

for all i then

$$\Pr[\bar{A}_1 \cdots \bar{A}_n] > \prod_{i=1}^{n} (1 - y_i \Pr[A_i]) > 0.$$

Lower Bound $R(3, k)$. Let us apply the general form of the Lovász Local Lemma to give a lower bound to $R(3, k)$. Recall that the basic probabilistic method gave "nothing" and the deletion method gave $R(3, k) > k^{3/2+o(1)}$. Color the edges of K_n independently with each edge Red with probability p. For each 3-set S let A_S be "S is Red" and for each k-set T let B_T be "T is Blue." Then

$$\Pr[A_S] = p^3, \quad \Pr[B_T] = (1-p)^{\binom{k}{2}} \sim e^{-pk^2/2}.$$

Let S, S' be adjacent in the dependency graph if they have a common edge; the same for S, T or T, T'. Each S is adjacent to only $3(n-3) \sim 3n$ other S' (a critical savings). Each T is adjacent to less than $\binom{k}{2}n < k^2n/2$ of the S. We use only that each S or T is adjacent to at most $\binom{n}{k}$—that is, all—of the T. Suppose that with each A_S we associate the same $y_S = y$, and with each B_T the same $y_T = z$. Then the Lovász Local Lemma takes roughly the following form:

If there exist p, y, z with

$$\ln y > y(3n)p^3 + z\binom{n}{k} e^{-pk^2/2},$$

$$\ln z > y(k^2n/2)p^3 + z\binom{n}{k} e^{-pk^2/2},$$

$$yp^3 < 1, \qquad z e^{-pk^2/2} < 1,$$

then $R(3, k) > n$.

What is the largest $k = k(n)$ so that p, y, z exist satisfying these conditions? Elementary analysis (and a free weekend!) give that the optimum is achieved at

$$p = c_1 n^{-1/2}, \qquad k = c_2 n^{1/2} \ln n,$$
$$z = \exp[c_3 n^{1/2} (\ln n)^2], \qquad y = 1 + \varepsilon,$$

with appropriate values for the constants. If we express n in terms of k

$$R(3, k) > ck^2/\ln^2 k.$$

This matches a 1961 result of Erdös in which a highly subtle deletion method was used.

What about $R(4, k)$? The basic probabilistic method gave a lower bound $k^{3/2+o(1)}$. The deletion method improved this to $k^{2+o(1)}$. From the Lovász Local Lemma one may obtain (try it!) $R(4, k) > k^{5/2+o(1)}$. The upper bound is $R(4, k) < k^{3+o(1)}$, so a substantial gap still remains.

REFERENCES

The Lovász Local Lemma first appeared in:

P. ERDÖS AND L. LOVÁSZ, *Problems and results on 3-chromatic hypergraphs and some related questions*, in Infinite and Finite Sets, North Holland, Amsterdam–New York, 1975.

The applications to the Ramsey function are in:

J. SPENCER, *Asymptotic lower bounds for Ramsey functions*, Discrete Math., 20 (1977), pp. 69–76.

The algorithmic implementation of the Lovász Local Lemma is in:

J. BECK, *An algorithmic approach to the Lovász Local Lemma*, I., Random Structures & Algorithms, 2 (1991), pp. 343–366.

and

N. ALON, *A parallel algorithmic version of the Local Lemma*, Random Structures & Algorithms, 2 (1991), pp. 367–378.

LECTURE 9

Discrepancy II

For a family of sets $\mathcal{A} \subset 2^{\Omega}$ the degree $\deg(x)$ is the number of $S \in \mathcal{A}$ with $x \in S$ and the degree $\deg(\mathcal{A})$ is the maximal $\deg(x)$ over $x \in \Omega$. The main result of this lecture is the following theorem.

The Beck–Fiala theorem. *If* $\deg(\mathcal{A}) \le t$, *then* $\operatorname{disc}(\mathcal{A}) \le 2t-1$.

It is remarkable to me that any bound on $\operatorname{disc}(\mathcal{A})$ may be made when the number of sets and the size of the sets are unbounded, and only the degree is bounded. Still, we will explore the possibility of an even stronger result.

CONJECTURE. *If* $\deg(\mathcal{A}) \le t$, *then* $\operatorname{disc}(\mathcal{A}) \le Kt^{1/2}$.

We will first give a result weaker than the Beck-Fiala Theorem with the hope that the methods in it can perhaps be improved.

WEAK THEOREM. *If* $\deg(\mathcal{A}) \le t$, *then* $\operatorname{disc}(\mathcal{A}) \le t \lg t(1+o(1))$.

The Pigeonhole Principle. By a partial coloring of Ω we mean a map $\chi : \Omega \to \{-1, 0, 1\}$ with $\chi \ne 0$. We think of χ as a two-coloring with $\chi(a)=0$, meaning that a has not been colored. We define $\chi(S) = \sum_{a \in S} \chi(a)$ as before. A partial coloring χ is called perfect on $\mathcal{A}$ if $\chi(A)=0$ for all $A \in \mathcal{A}$.

THEOREM. *If* $\mathcal{A} \subset 2^{\Omega}$ *and*

$$\prod_{S \in \mathcal{A}} (1+|S|) < 2^{|\Omega|},$$

then there exists a perfect partial coloring.

Proof. Write $\mathcal{A} = \{S_1, \cdots, S_m\}$ for convenience. With every $\chi : \Omega \to \{-1, +1\}$ we associate the m-tuple

$$\Psi(\chi) = (\chi(S_1), \cdots, \chi(S_m)).$$

There are at most $\prod (1+|S_i|)$ possible values for $\Psi(\chi)$ since the ith coordinate is one of $1+|S_i|$ values. The Pigeonhole Principle says that χ is not injective. There exist χ_1, χ_2 with $\Psi(\chi_1) = \Psi(\chi_2)$, i.e., $\chi(S_i) = \chi_2(S_i)$ for each i. Now set

$$\chi = (\chi_1 - \chi_2)/2.$$

Observe that $\chi(i) \in \{-1, 0, 1\}$ with $\chi(i) = 0$ if and only if $\chi_1(i) = \chi_2(i)$. For all i

$$\chi(S_i) = (\chi_1(S_i) - \chi_2(S_i))/2 = 0$$

so χ is a perfect partial coloring. □

In the final two lectures we will see that the Pigeonhole Principle is a powerful tool that can sometimes be effectively melded with probabilistic techniques. The Pigeonhole Principle is essentially existential in nature—more so, I feel, than the probabilistic method. For example, any family of $[n/\log_2 11]$ 10-sets on $[n]$ has a perfect partial coloring but no polynomial time algorithm is known to find it.

The weak theorem. *Fix* $R \geq t \geq 3$ *and let* $S_1, \cdots, S_m \subset [n]$ *with* $\deg(j) \leq t$ *for all* j *and* $|S_i| \geq R$ *for all* i. *Doublecounting*

$$mR < \sum |S_i| = \sum \deg(j) < nt$$

so

$$m < n(t/R).$$

Now

$$\sum |S_i| + 1 \leq nt + m \leq n(t+1).$$

Fixing a sum and a lower bound the product is maximized basically when all terms are as small as possible. More precisely, we have the following lemma.

LEMMA. *If* $y_1 + \cdots + y_m = A$, $y_i \geq K$ *for all* i, $K \geq e$, *then*

$$\prod y_i < K^{A/K}.$$

Proof.

$$\ln\left(\prod y_i\right) = \sum \ln y_i = \sum y_i[(\ln y_i)/y_i].$$

The function $f(y) = (\ln y)/y$ is decreasing for $y \geq e$. Hence all $(\ln y_i)/y_i \leq (\ln K)/K$. Thus

$$\ln\left(\prod y_i\right) \leq \sum y_i[(\ln K)/K] = A(\ln K)/K$$

and we get the lemma by exponentiating both sides.

Let us assume now that $(R+1)^{t+1} < 2^{R+1}$. Then if we apply the lemma

$$\prod (1 + |S_i|) \leq (R+1)^{n(t+1)/(R+1)} < 2^n,$$

so there would exist a perfect partial coloring of $\mathcal{A}$. □

THEOREM. *Let* R, t *satisfy* $(R+1)^{t+1} < 2^{R+1}$. *If* $\deg(\mathcal{A}) \leq t$, *then* $\operatorname{disc}(\mathcal{A}) \leq R$. Asymptotically we may take $R = t \lg t(1 + o(1))$.

Proof. Let $\mathcal{A} = \mathcal{A}_0$ be an arbitrary family on $\Omega = \Omega_0$ with $\deg(\mathcal{A}) \leq t$. Set $\mathcal{A}_0^* = \{A \in \mathcal{A}_0: |A| \geq R\}$ and let χ_0 be a perfect partial coloring of $\mathcal{A}_0^*$. (That is, ignore the small sets.) Set $\Omega_1 = \{a \in \Omega_0: \chi_0(a) = 0\}$ (the uncolored points), $\mathcal{A}_1 = \mathcal{A}_0|_{\Omega_1}$ and iterate. We find a sequence $\chi_0, \chi_1, \cdots$ of partial colorings which terminates when $\Omega_s = \emptyset$. This defines a coloring χ by $\chi(a) = \chi_i(a)$, where $a \in \Omega_i - \Omega_{i+1}$. For any $A \in \mathcal{A}$ let r be the first integer such that $|A \cap \Omega_r| < R$. For

$0 \le i < r$, $A \in \mathcal{A}_i^*$ and so $\chi_i(A) = 0$. Then

$$\begin{aligned}\chi(A) &= \sum_{i=0}^{s-1} \chi_i(A) \\ &= \sum_{i=0}^{r-1} \chi_i(A) + \chi(A \cap \Omega_r) \\ &= 0 + \chi(A \cap \Omega_r) \\ &\le |A \cap \Omega_r| < R.\end{aligned}$$

In other words, A is colored perfectly until it has fewer than R elements uncolored. Then it is ignored, but its discrepancy can become no larger than $R-1$. □

Floating colors. A colorating is a map $\chi: \Omega \to \{-1, +1\}$. The key to the Beck-Fiala Theorem is to consider the values $\chi(a)$ as variables lying anywhere in $[-1, +1]$. Initially all $\chi(a)$ are set equal to zero. All sets then have zero discrepancy. At the end all $\chi(a)$ must be -1 or $+1$. We describe the procedure that is to be iterated to go from the initial trivial coloration to the final "real" coloration.

For convenience let $S_1, \cdots, S_m$ be the sets and $1, \cdots, n$ the points of the system. Suppose values $p_1, \cdots, p_n \in [-1, +1]$ are given. (p_j is the "current" value of $\chi(j)$.) Call j constant if $p_j = \pm 1$; otherwise call j floating. Call S_i ignored if it has at most t floating points; otherwise call S_i active. Assume that all active sets have zero sum, i.e.,

$$\sum_{j \in S_i} p_j = 0, \qquad S_i \text{ active}.$$

We want to move the floating p_j so that the active sets still have zero sum and some floating p_j becomes constant. Renumber for convenience so that $1, \cdots, s$ are the floating points and $S_1, \cdots, S_r$ are the active sets. Each point is in at most t sets altogether, and each active set contains more than t floating points so $r < s$. Let $(y_1, \cdots, y_s)$ satisfy

$$\sum_{j \in S_i} y_j = 0, \qquad 1 \le i \le r.$$

As $r < s$ there is a nonzero solution to this system. Let λ be the minimal positive real so that some $p_j + \lambda y_j \in \{-1, +1\}$. Set

$$\begin{aligned}p'_j &= p_j + \lambda y_j, \qquad 1 \le j \le s, \\ &= p_j, \qquad j > s.\end{aligned}$$

Any active set S has

$$\sum_{j \in S'} p'_j = \sum_{j \in S} p_j + \lambda \sum_{j \in S} y_j = 0 + \lambda(0) = 0.$$

The choice of λ insures that all $p'_j \in [-1, +1]$ and the $p'_1, \cdots, p'_n$ have fewer floating points than $p_1, \cdots, p_n$.

BECK-FIALA THEOREM. *If* $\deg(\mathscr{A}) \le t$, *then* $\mathrm{disc}(\mathscr{A}) \le 2t-1$.

Proof. Set $p_1 = \cdots = p_n = 0$ initially and iterate the above procedure until finding final $p_1^F, \cdots, p_n^F$. Let $S \in \mathscr{A}$ and let $p_1, \cdots, p_n$ be the values of the p_i when S first becomes ignored. At that stage $\sum_{j \in S} p_j = 0$. Thus

$$\left|\sum_{j \in S} p_j^F\right| = \left|\sum_{j \in S} p_j^F - p_j\right| \le \sum_{j \in S} |p_j^F - p_j|.$$

When S becomes ignored it has at most t floating points. The constant p_j change nevermore so $p_j = p_j^F$. The floating points must do so within the interval $[-1, +1]$ so $|p_j - p_j^F| \le 2$. Hence

$$\left|\sum_{j \in S} p_j^F\right| \le 2t.$$

We have shown $\mathrm{disc}(\mathscr{A}) \le 2t$. The improvement to $\mathrm{disc}(\mathscr{A}) \le 2t-1$ comes (exercise) by choosing λ with minimal absolute value such that some $p_j + \lambda y_j = \pm 1$. □

The Komlos conjecture. The Beck–Fiala Theorem can be easily rewritten in vector form. For $x = (x_1, \cdots, x_m) \in R^m$ as notation for the L^1, L^2, and L^∞ norms we use

$$|x|_1 = \sum |x_i|, \quad |x|_2 = [\sum x_i^2]^{1/2}, \quad |x|_\infty = \max |x_0|.$$

BECK-FIALA THEOREM (Vector Form). *Let* $u_1, \cdots, u_n \in R^m$ *with all* $|u_j|_1 \le 1$. *Then for some choice of signs*

$$|\pm u_1 \pm \cdots \pm u_n|_\infty \le 2.$$

Proof. Let $u_j = (a_{1j}, \cdots, a_{mj})$, the column vectors of an $m \times n$ matrix $A = [a_{ij}]$. (In the 0-1 case A is the incidence matrix of the family $\mathscr{A}$.) Let $p_1, \cdots, p_m \in [-1, +1]$ be given. Call p_j fixed if $p_j = \pm 1$; otherwise floating. Call row i ignored if $\sum' |a_{ij}| \le 1$, the sum over the floating j; otherwise call row i active. As each column has the sum of its absolute values at most one (the L^1 condition) there are fewer active rows than floating columns. Find y_j, over the floating j, with $\sum a_{ij} y_j = 0$ for each active row i. As this system is underdetermined there is a nonzero solution. Now replace p_j with $p_j + \lambda y_j$ with λ so that all p_j remain in $[-1, +1]$ and some p_j becomes constant.

Initially we set all $p_j = 0$ and iterate the above procedure until all $p_j = \pm 1$. A given row has zero sum (i.e. $\sum a_{ij} p_j$) until it becomes ignored. Then each p_j changes by at most two, so the sum also changes by at most two. □

Here is an extremely interesting problem to which I have devoted a great deal of effort.

KOMLOS CONJECTURE. *There is an absolute constant K so that for all n, m and all* $u_1, \cdots, u_n \in R^m$ *with all* $|u_j|_2 \le 1$ *there exist signs* $\pm$ *so that*

$$|\pm u_1 \pm \cdots \pm u_n|_\infty \le K.$$

By the reductions of Lecture 5 it would suffice to prove this conjecture when $n = m$. Let the u_j be the column vectors of $A = [a_{ij}]$ of size $n \times n$. Then $\sum_j \sum_i a_{ij}^2 \le \sum_j 1 = n$. Set $\sigma_i = [\sum_j a_{ij}^2]^{1/2}$ so $\sum_i \sigma_i^2 = n$. If all the σ_i were roughly one (but perhaps they are not) a random choice of $\pm$ would give $\pm u_1 \pm \cdots \pm u_n = (L_1, \cdots, L_n)$ with the L_i roughly standard normal. We would want all $|L_i| \le K$. The ideas of the final lecture give us hope, but no proof.

Let $\mathcal{A} \subseteq 2^{\Omega}$ with $\deg(\mathcal{A}) \le t$. The incidence matrix A has column vectors u_j with $|u_j|_2 \le t^{1/2}$. The Komlos Conjecture would imply the existence of signs with $|u_1 \pm \cdots \pm u_n|_\infty \le Kt^{1/2}$. The signs are a coloring χ, $|\pm u_1 \pm \cdots \pm u_n|_\infty = \operatorname{disc}(\mathcal{A}, \chi)$ so $\operatorname{disc}(\mathcal{A}) \le Kt^{1/2}$. That is, the Komlos Conjecture would imply the conjectured improvement on the Beck–Fiala Theorem.

Most problems involving balancing of vectors use a single norm to bound both the vectors and their sum. It is intriguing that both the Beck–Fiala Theorem and the Komlos Conjecture are exceptional in this respect.

The Barany–Grunberg Theorem. *Let $v_1, v_2, \cdots$ be an infinite sequence of vectors in R^n with $|v_i| \le 1$. Then there exist signs $\varepsilon_i = \pm 1$ so that*

$$\left|\sum_{i=1}^{t} \varepsilon_i v_i\right| \le 2n, \qquad t = 1, 2, \cdots.$$

Proof. Suppose $p_1, \cdots, p_t \in [-1, +1]$ are such that
(i) at most n of the p_i are floating (i.e. in $(-1, +1)$),
(ii) $p_1 v_1 + \cdots + p_t v_t = 0$.
We give a procedure to go from t to $t+1$. Introduce $p_{t+1} = 0$. Then (ii) is certainly satisfied but perhaps $n+1$ of the p_i are floating. The equation $\sum y_j v_j = 0$, summation over the floating j, is underdetermined and so it has a nonzero solution. Let λ be the minimal positive real so that some floating j has $p_j + \lambda y_j = \pm 1$. Then reset p_j to $p_j + \lambda y_j$ for each floating j. Condition (ii) holds as

$$\sum (p_j + \lambda y_j) = \sum p_j v_j + \lambda \sum y_j v_j = 0 + 0 = 0$$

and some floating j has been made constant, so (i) holds.

Begin the procedure at "time" $t = n$ with $p_1 = \cdots = p_n = 0$. Continue it "forever." If p_i ever becomes ± 1 it stays there; let p_i^F be that value. Otherwise set $p_i^F = \pm 1$ arbitrarily. For a given t let $p_1, \cdots, p_t$ be the values at "time" t so that $p_1 v_1 + \cdots + p_t v_t = 0$. Then

$$|p_1^F v_1 + \cdots + p_t^F v_t| = \left|\sum_{i=1}^{t} (p_i^F - p_i) v_i\right|$$

$$\le \sum_{i=1}^{t} |p_i^F - p_i||v_i|.$$

Now $p_i^F = p_i$ for all but at most n terms. For these terms $|p_i^F - p_i| \le 2$ and $|v_i| \le 1$ so their product is at most two. Thus $|p_1^F v_1 + \cdots + p_t^F v_t| \le 2n$. $\square$

Have we forgotten something? What norm are we using? That is the amazing part. The Barany–Grunberg Theorem applies for *any* norm on R^n! This leads to a nice question: Can we improve this result in, say, the L^2 norm? It is conjectured that if $u_i \in R^n$ with all $|u_i|_2 \le 1$, then there exist $\varepsilon_i = \pm 1$ so that

$$\left|\sum_{i=1}^{t} \varepsilon_i u_i\right|_2 < n^{1/2+o(1)} \qquad \text{for all } t.$$

An olympiad problem. The following problem appeared on the 1986 International Mathematical Olympiad, a contest between national teams of high school students. Try it!

> Let S be a finite set of lattice points. Prove that there exists a coloring of S Red and Blue so that in every horizontal and vertical line the number of Red and Blue points are either equal or differ by one.

Let $\mathcal{A}$ be the family of horizontal and vertical lines of S. In our language the problem asks us to prove disc $(\mathcal{A}) \leq 1$. Each point is in two lines, the horizontal and the vertical, so deg $(\mathcal{A}) \leq 2$. The Beck–Fiala Theorem then gives disc $(\mathcal{A}) \leq 3$. Cecil Rousseau, coach of the U.S. team, estimates that this result would have received half credit.

Here is the solution: Suppose there is a "cycle" $P_0P_1 \cdots P_{2k-1}$ with P_iP_{i+1} on a common line which alternates horizontal and vertical with i, including $P_{2k-1}P_0$. Then coloring P_{2i} Red and P_{2i+1} Blue gives a perfect partial coloring and the result follows by induction. Otherwise let $P_0 \cdots P_j$ be a maximal sequence with P_iP_{i+1} on a common line alternating horizontal and vertical. Again color P_{2i} Red and P_{2i+1} Blue. All lines are perfectly balanced, except for the one through P_0 not in the P_1 direction and the analogous one through P_j. These lines have imbalance one, but by maximality there are no other points on these lines other than the P_i. Therefore the result again follows by induction.

Work in progress. Here is an attempt at improving the Beck–Fiala Theorem by showing disc $(\mathcal{A}) < t^{1/2+\varepsilon}$. The attempt has not (yet) been successful, and the crude estimates we give reflect the mindset in which one tries to improve a result. Let Ω have n elements and let $\mathcal{A}$ be a family with degree t. Call S large if $|S| > 100t \ln t$. Call S small if $|S| < t^{1/2+\varepsilon}$. Call S medium if $t^{1/2+\varepsilon} < |S| < 100t \ln t$. We ignore the small sets and find a partial coloring perfect on the large sets and reasonable on the medium sets. Let $\mathcal{C}$ be the set of colorings $\chi : \Omega \to \{-1, +1\}$ such that

$$(*) \qquad |\chi(S)| < 100|S|^{1/2}(\ln t)^{1/2}$$

for all median S. To estimate $\mathcal{C}$ let χ be random and A_S the event that $(*)$ fails. Then (Lecture 4) $\Pr[A_S] < t^{-5000}$. Each A_S is adjacent in the dependency graph to at most $|S|t < t^2$ other $A_{S'}$ as deg $(\mathcal{A}) < t$. The Lovász Local Lemma applies and

$$\Pr[\wedge \bar{A}_S] > \prod (1 - 2\Pr[A_S]) > (1 - 2t^{-5000})^{nt} > .99^n$$

and so $\mathcal{C} > 1.98^n$. Let $S_1, \cdots, S_r$ be the large sets and map $\chi \in \mathcal{C}$ into $\psi(\chi) = (\chi(S_1), \cdots, \chi(S_r))$. There are 1.98^n pigeons and fewer than 1.01^n pigeonholes; therefore there is a $\mathcal{C}' \subset \mathcal{C}$, $|\mathcal{C}'| > 1.96^n$ on which $\chi(S)$ is constant for all large S. Fix $\chi_1 \in \mathcal{C}'$. The number of $\chi \in \mathcal{C}$ that differ from χ_1 in at most $.4n$ positions is

$$\sum_{i=0}^{.4n} \binom{n}{i} < n\binom{n}{.4n} \ll 1.96^n.$$

Thus some $\chi_2 \in \mathcal{C}'$ has $\chi_2(i) \neq \chi_1(i)$ for at least $.4n$ i's. Now set $\chi = (\chi_1 - \chi_2)/2$. χ is a partial coloring that

(i) is perfect on the large sets;
(ii) is reasonable—$|\chi(S)| < 100|S|^{1/2}(\ln t)^{1/2}$—on the medium sets;
(iii) colors at least 40% of the points.

Now iterate this procedure until all points have been colored.

What can go wrong? The large sets are being colored perfectly, the small sets can do limited harm; the problems arise with a set of medium size. While $\chi(S) < 100|S|^{1/2}(\ln t)^{1/2}$ and 40% of the points overall are colored, we do not know how many points of S are colored. Maybe only some $k < 100|S|^{1/2}(\ln t)^{1/2}$ points of S are colored and they are all colored Red! This seems "unlikely." Furthermore if 40% of the points are colored, then most medium S should have many of their points colored. There are many ways to choose χ_2 and therefore χ; can we not pick one that does well in the sense of intersecting the medium sets? But this is all speculation and frustration—there is no proof.

If we bound n then (iii) helps us. The procedure is only iterated $c \ln n$ times. We do get the following result of Jozsef Beck.

THEOREM. *If* $\mathcal{A} \subseteq 2^\Omega$ *has* $\deg(\mathcal{A}) \leq t$ *and* $|\Omega| = n$, *then*

$$\operatorname{disc}(\mathcal{A}) < ct^{1/2}(\ln t)^{1/2}(\ln n).$$

Notice that if the Beck-Fiala Theorem is the best possible, then an example showing $\operatorname{disc}(\mathcal{A}) > ct$ will have to be very large—$n > e^{t^{1/2-o(1)}}$—indeed.

REFERENCES

For the Beck-Fiala Theorem:

J. BECK AND T. FIALA, *Integer-making theorems*, Discrete Appl. Math., 3 (1981), pp. 1-8.

For the Barany-Grunberg Theorem:

I. BARANY AND V. S. GRUNBERG, *On some combinatorial questions in finite dimensional spaces*, Linear Algebra Appl., 41 (1981), pp. 1-9.

Partial results on the Komlos Conjecture are given in my paper cited in Lecture 10.

LECTURE 10

Six Standard Deviations Suffice

I suppose every mathematician has a result he is most pleased with. This is mine.

THEOREM. *Let* $S_1, \cdots, S_n \subset [n]$. *Then there exists* $\chi:[n]\to\{-1,+1\}$ *such that*

$$|\chi(S_i)| < 6n^{1/2}$$

for *all* i, $1 \le i \le n$.

The elementary methods of Lecture 1 already give χ with all $|\chi(S_i)| < cn^{1/2}(\ln n)^{1/2}$. From our vantage point $n^{1/2}$ is one standard deviation. With χ random $|\chi(S)| > 6n^{1/2}$ occurs with a small but positive probability $\varepsilon < e^{-6^2/2} = e^{-18}$. There are n sets, so the expected number of i with $|\chi(S_i)| > 6n^{1/2}$ is εn, which goes to infinity with n. A random χ will not work; the key is to meld probabilistic ideas with the Pigeonhole Principle.

The constant 6 is simply the result of calculation, the essential point is that it is an absolute constant. In the original paper it is 5.32, in our proof here "6" = 11.

Proof. Let C be the set of $\chi:[n]\to\{-1,+1\}$. Call $\chi \in C$ *realistic* if

(1) $|\chi(S_i)| > 10n^{1/2}$ for at most $4(2e^{-50})n$ i's,
(2) $|\chi(S_i)| > 30n^{1/2}$ for at most $8(2e^{-450})n$ i's,
(3) $|\chi(S_i)| > 50n^{1/2}$ for at most $16(2e^{-1250})n$ i's,

and, in general,

(s) $|\chi(S_i)| > 10(2s-1)n^{1/2}$ for at most $2^{s+1}(2e^{-50(2s-1)^2})n$ i's.

CLAIM. At least half the $\chi \in C$ are realistic.

Pick $\chi \in C$ at random. Let Y_i be the indicator random variable for $|\chi(S_i)| > 10n^{1/2}$. Set $Y = \sum_{i=1}^{n} Y_i$. Then

$$E[Y_i] = \Pr[|\chi(S_i)| > 10n^{1/2}] < 2e^{-50}$$

by the bound of Lecture 4. By linearity of expectation

$$E[Y] = \sum_{i=1}^{n} E[Y_i] = (2e^{-50})n.$$

(We do not know much about the distribution of Y since the dependence of the Y_i may be complex, reflecting the intersection pattern of the S_i. Fortunately, linearity of expectation ignores dependency.) As $Y \geq 0$

$$\Pr[Y > 4E[Y]] < \tfrac{1}{4}.$$

That is,

$$\Pr[\chi \text{ fails } (1)] < \tfrac{1}{4}.$$

Apply the same argument to (2), letting Y_i be one if $|\chi(S_i)| > 30n^{1/2}$. The $10^2/2 = 50$ becomes $30^2/2 = 450$. Everything is identical except that 4 was changed to 8 so that

$$\Pr[\chi \text{ fails } (2)] < \tfrac{1}{8}.$$

In general,

$$\Pr[\chi \text{ fails } (s)] < 2^{-s-1}.$$

The probability of a disjunction is at most the sum of the probabilities.

$$\Pr[\chi \text{ not realistic}] \leq \sum_{s=1}^{\infty} 2^{-s-1} = \tfrac{1}{2}.$$

A random χ has probability at least $\frac{1}{2}$ of being realistic. Each χ was given equal weight 2^{-n} in the probability space so at least 2^{n-1} χ are realistic, completing the claim.

Now we define a map

$$T(\chi) = (b_1, \cdots, b_n)$$

where b_i is the nearest integer to $\chi(S_i)/20n^{1/2}$. That is,

-2	-1	0	$+1$	$+2$	b_i
$-30\sqrt{n}$	$-10\sqrt{n}$ $\quad 0$	$10\sqrt{n}$	$30\sqrt{n}$	$50\sqrt{n}$	$\chi(S_i)$

$b_i = 0$ means $|\chi(S_i)| \leq 10n^{1/2}$,

$b_i = 0, \pm 1$ means $|\chi(S_i)| < 30n^{1/2}$, etc.

Let B be the set of possible $T(\chi)$ with χ realistic. That is, B is all $(b_1, \cdots, b_n)$ such that

(1) $b_i \neq 0$ for at most $4(2e^{-50})n$ i's,

(2) $b_i \neq 0, \pm 1$ for at most $8(2e^{-450})n$ i's,

and, in general,

(s) $|b_i| \geq s$ for at most $2^{s+1}(2e^{-50(2s-1)^2})n$ i's.

(Let us pause for a geometric view. We are actually showing that if $A = [a_{ij}]$ is an $n \times n$ matrix with all $|a_{ij}| \leq 1$, then there exists $x \in C = \{-1, +1\}^n$ so that $|Ax|_\infty \leq Kn^{1/2}$. We consider A: $R^n \to R^n$ given by $x \to Ax$. We split the range into cubes of size $20n^{1/2}$, centered about the origin. We want to prove that some $x \in C$ is mapped into the central cube. We cannot do this directly, but we plan to find $x, y \in C$ mapped into the same cube and examine $(x-y)/2$. We want to employ the Pigeonhole Principle, with the 2^n $x \in C$ mapped into the cubes. To do this we restrict attention to the 2^{n-1} realistic x as they are mapped (shown by the next claim) into a much smaller number of cubes.)

CLAIM. $|B| < (1.0000000000000001)^n$.

We use the inequality, valid for all n, all $a \in [0, 1]$,

$$\sum_{i<na} \binom{n}{i} < 2^{nH(a)}$$

where $H(a) = -a \lg a - (1-a) \lg (1-a)$ is the entropy function. We can choose $\{i: b_i \neq 0\}$ in at most $2^{nH(8e^{-50})}$ ways. Then we can choose the signs of the nonzero b_i in at most $2^{8e^{-50}n}$ ways. For each s there are at most $2^{**}[nH(2^{s+1}e^{-50(2s+1)^2})]$ ways to choose $\{i: |b_i| > s\}$. These choices determine $(b_1, \cdots, b_n)$. Thus $|B| < 2^{\beta n}$ where

$$\beta = 8e^{-50} + H(8e^{-50}) + H(16e^{-450}) + H(32e^{-1250}) + \cdots.$$

This series clearly converges and the claim follows from a calculation.

Now apply the Pigeonhole Principle to the map T from the (at least) 2^{n-1} realistic χ to the (at most) $(1+10^{-16})^n$ pigeonholes B. There is a set C' of at least $2^{n-1}/(1+10^{-16})^n$ realistic χ mapping into the same $(b_1, \cdots, b_n)$.

Let us think of C as the Hamming Cube $\{-1, +1\}^n$ with the metric

$$\rho(\chi_1, \chi_2) = |\{i: \chi_1(i) \neq \chi_2(i)\}|.$$

D. Kleitman has proved that if $C' \subset C$ and $C' \geq \sum_{i \leq r} \binom{n}{i}$ with $r \leq n$ then C' has diameter at least $2r$. That is, the set of a given size with minimal diameter is the ball. In our case $|C'| > 2^{n-1}/(1+10^{-16})^n$ so we may take $r = \frac{1}{2}n(1-10^{-6})$ with room to spare and C' has diameter at least $n(1-10^{-6})$. Let $\chi_1, \chi_2 \in C'$ be at maximal distance. (Kleitman's Theorem is not really necessary. The elementary argument at the end of Lecture 9 gives that C' has diameter at least $.4999n$. We could use this value and finish the proof with a much worse, but still absolute constant, value of "6".) Now set

$$\chi = (\chi_1 - \chi_2)/2;$$

then χ is a partial coloring of $[n]$. As $T(\chi_1) = T(\chi_2)$ both $\chi_1(S_i)$ and $\chi_2(S_i)$ lie in a common interval $[(20b_i - 10)n^{1/2}, (20b_i + 10)n^{1/2}]$. Then

$$(*) \qquad |\chi(S_i)| = |(\chi_1(S_i) - \chi_2(S_i))/2| \leq 10n^{1/2}.$$

Also

$$(**) \qquad |\{i: \chi(i) \neq 0\}| < 10^{-6}n.$$

Now iterate. We now have n sets on $10^{-6}n$ points. If we had only $10^{-6}n$ sets we could partially color all but a millionth of the points, giving all sets discrepancy of at most $10(10^{-6}n)^{1/2} = .01n^{1/2}$. Things are not quite so simple as we still have n sets. We actually need the following result: Given n sets on r points, $r \leq n$, there is a partial coloring of all but at most a millionth of the points so that all

$$|\chi(S)| < 10r^{1/2}[\ln(2n/r)]^{1/2}.$$

The argument is basically that given when $r = n$ but the formulae are a bit more fierce. Let us assume the result (read the original paper!) and now iterate. On the second iteration

$$|\chi(S)| < 10(n10^{-6})^{1/2}[\ln(2\times 10^6)]^{1/2} < .4n.$$

The future terms decrease even faster. The logarithmic term, while annoying, does not affect the convergence. At the end, with all points colored,

$$\begin{aligned}|\chi(S)| &\leq 10n^{1/2} + 10(n10^{-6})^{1/2}[\ln(2\times 10^6)]^{1/2}\\ &\quad + 10(n10^{-12})^{1/2}[\ln(2\times 10^{12})]^{1/2}\\ &\quad + \cdots\\ &\leq 11n^{1/2},\end{aligned}$$

completing the proof for "6" = 11.

From the reductions of Lecture 5 we derive the following.

COROLLARY. $\operatorname{disc}(\mathcal{A}) \leq K|\mathcal{A}|^{1/2}$. *That is, given any n finite sets there is a two-coloring of the underlying points so that each set has discrepancy at most* $Kn^{1/2}$.

This result is best possible up to the constant. Here are two proofs. First, take an $n\times n$ Hadamard Matrix $H = (h_{ij})$ with the first row all ones. Set $A = (a_{ij}) = (H+J)/2$ so that $a_{ij} = 1$ when $h_{ij} = 1$ and $a_{ij} = 0$ when $h_{ij} = -1$. Let $\vec{1} = v_1, v_2, \cdots, v_n$ be the columns of H and $\vec{1} = w_1, w_2, \cdots, w_n$ be the columns of A, so that $w_i = (v_i + \vec{1})/2$. For any choice of signs

$$u = \pm w_1 \pm w_2 \pm \cdots \pm w_n = \tfrac{1}{2}v + s\vec{1}$$

where $v = \pm v_2 \pm \cdots \pm v_n$. As the v_i are orthogonal and $|v_i|_2 = n^{1/2}$, $|v|_2 = [n(n-1)]^{1/2}$. Also $v\cdot\vec{1} = 0$ so $|u|_2 \geq \frac{1}{2}|v|_2 = [n(n-1)]^{1/2}/2$ and thus

$$|u|_\infty \geq (n-1)^{1/2}/2.$$

The second proof involves turning the probabilistic method on its head. Let $T_1, \cdots, T_n$ be randomly chosen subsets of $[n]$. That is, for all i, j $\Pr[j \in T_i] = \frac{1}{2}$ and these events are mutually independent. Let $\chi: [n] \to \{-1, +1\}$ be arbitrary but fixed. Let $P = \{j: \chi(j) = +1\}$, $N = \{j: \chi(j) = -1\}$, $a = |P|$ so $n - a = |N|$. Then $|T_i \cap P|$ has binomial distribution $B(a, \frac{1}{2})$ while $|T_i \cap N|$ has $B(n-a, \frac{1}{2})$ and thus $\chi(T_i)$ has distribution $B(a, \frac{1}{2}) - B(n-a, \frac{1}{2})$. When $a = n/2$ $\chi(T_i)$ is roughly normal with zero mean and standard deviation $\frac{1}{2}n^{1/2}$. Then

$$\lim_n \Pr[|\chi(T_i)| \leq \tfrac{1}{2}c\sqrt{n}] = \int_{-c}^{+c} \frac{1}{\sqrt{2\pi}} e^{-t^2/2}\, dt.$$

One can show that this probability is maximized essentially when $a = n/2$. Pick $c \sim .67$ so that the above integral is .5. Decrease c slightly so that the inequality is strict:

$$\Pr[|\chi(T_i)| < .33\sqrt{n}] < .499.$$

Now since the T_i are chosen independently the events $|\chi(T_i)|>.33n^{1/2}$ are mutually independent so that

$$\Pr[|\chi(T_i)|<.33n^{1/2}, 1\le i\le n]<.499^n.$$

Let Y_χ be one if $|\chi(T_i)|<.33n^{1/2}$ for all i and let

$$Y=\sum Y_\chi$$

the sum over all 2^n colorings χ. Then

$$E[Y]=\sum_\chi E[Y_\chi]<2^n(.499)^n \ll 1.$$

Thus the event $Y=0$ has positive probability, actually probability nearly one. There is a point in the probability space (i.e. actual sets $T_1,\cdots,T_n$) so that $Y=0$ which, when we unravel the definitions, means that the family $\mathcal{A}=\{T_1,\cdots,T_n\}$ has disc $(\mathcal{A})>.33n^{1/2}$.

Let us restate our basic theorem of this Lecture in vector form.

THEOREM. *Let $u_j\in R^n$, $1\le j\le n$, $|u_j|_\infty\le 1$. Then for some choice of signs*

$$|\pm u_1\pm\cdots\pm u_n|_\infty\le Kn^{1/2}.$$

To prove this we set $u=(L_1,\cdots,L_n)=\pm u_1\pm\cdots\pm u_n$. Then each L_i has distribution $L_i=\pm a_{i1}\pm\cdots\pm a_{in}$ with all $|a_{ij}|\le 1$. From the arguments of Lecture 4, $\Pr[|L_i|>10n^{1/2}]<e^{-50}$, etc. and the proof goes through as before.

The methods of Lecture 5 also allow us to re-express our result in terms of simultaneous approximation. Given data a_{ij}, $1\le i\le m$, $1\le j\le n$ with all $a_{ij}\le 1$. Given initial values x_j, $1\le j\le n$. A simultaneous round-off is a set of integers y_j, each y_j either the "round-up" or "round-down" of x_j. Let E_i be the error

$$E_i=\sum_{j=1}^n a_{ij}(x_j-y_j)$$

and E the maximal error, $E=\max|E_i|$.

COROLLARY. *There exists a simultaneous round-off with $E\le Km^{1/2}$.*

Is there a polynomial time algorithm that gives the simultaneous round-off guaranteed by this corollary? Given $u_1,\cdots,u_n\in R^n$ with all $|u_j|_\infty\le 1$, is there a polynomial time algorithm to find signs such that $|\pm u_1\pm\cdots\pm u_n|_\infty<Kn^{1/2}$? The difficulties in converting these theorems to algorithms go back to the basic theorem of this Lecture and lie, I feel, in the use of the Pigeonhole Principle. In Lecture 4 we saw that there is a rapid algorithm giving $|\pm u_1\pm\cdots\pm u_n|_\infty<cn^{1/2}(\ln n)^{1/2}$. We also saw that no nonanticipative algorithm could do better. That is, a better algorithm could not determine the sign of u_j simply on the basis of $u_1,\cdots,u_{j-1},u_j$ but would have to look ahead. Also we can show, standing the probabilistic method on its head, that there are $u_1,\cdots,u_n$ so that the number of choices of signs with $|\pm u_1\pm\cdots\pm u_n|_\infty<Kn^{1/2}$ is less than $(2-c)^n$ of the 2^n possible choices. Hence a randomly selected choice of signs will not work. Let us rephrase back into the language of sets and conclude the Lectures with the following problem.

Open Problem. Is there a polynomial time algorithm which, given input $S_1, \cdots, S_n \subset [n]$, outputs a two-coloring $\chi: [n] \to \{-1, +1\}$ such that

$$|\chi(S_i)| \le Kn^{1/2}$$

for *all* i, $1 \le i \le n$?

REFERENCE

J. SPENCER, *Six standard deviations suffice*, Trans. Amer. Math. Soc., 289 (1985), pp. 679-706.

BONUS LECTURE

The Janson Inequalities

The inequalities. Let $A_1, \ldots, A_m$ be events in a probability space. Set

$$M = \prod_{i=1}^{m} \Pr[\bar{A}_i].$$

The Janson Inequality allows us, sometimes, to estimate $\Pr[\wedge \bar{A}_i]$ by M, the probability if the A_i were mutually independent. We let G be a dependency graph for the events in the sense of Lecture 8—i.e., the vertices are the indices $i \in [m]$ and each A_i is mutually independent of all A_j with j not adjacent to i in G. We write $i \sim j$ when i, j are adjacent in G. We set

$$\Delta = \sum_{i \sim j} \Pr[A_i \wedge A_j].$$

We make the following *correlation assumptions*:
(a) For all i, S with $i \notin S$

$$\Pr[A_i \mid \wedge_{j \in S} \bar{A}_j] \leq \Pr[A_i].$$

(b) For all i, k, S with $i, k \notin S$

$$\Pr[A_i \wedge A_k \mid \wedge_{j \in S} \bar{A}_j] \leq \Pr[A_i \wedge A_k].$$

Finally, let ε be such that $\Pr[A_i] \leq \varepsilon$ for all i.

The Janson Inequality. Under the above assumptions

$$M \leq \Pr[\wedge \bar{A}_i] \leq M e^{\Delta/[2(1-\varepsilon)]}.$$

We set

$$\mu = \sum \Pr[A_i],$$

the expected number of A_i that occur. As $1 - x \leq e^{-x}$ for all $x \geq 0$ we may bound $M \leq e^{-\mu}$ and then rewrite the upper bound in the somewhat weaker but quite convenient form

$$\Pr[\wedge \bar{A}_i] \leq e^{-\mu} e^{\Delta/[2(1-\varepsilon)]}.$$

In all our applications $\varepsilon = o(1)$ and the pesky factor of $1 - \varepsilon$ is no real trouble. Indeed just assuming all $\Pr[A_i] \le \frac{1}{2}$ is plenty for all cases we know of. In many cases we also have $\Delta = o(1)$. Then the Janson Inequality gives an asymptotic formula for $\Pr[\wedge \bar{A}_i]$. When $\Delta \gg \mu$, as also occurs in some important cases, the above gives an upper bound for $\Pr[\wedge \bar{A}_i]$ which is bigger than one. In those cases we sometimes can use the following:

The Extended Janson Inequality. Under the assumptions of the Janson Inequality and the additional assumption that $\Delta \ge \mu(1 - \varepsilon)$

$$\Pr[\wedge \bar{A}_i] \le e^{-\mu^2(1-\varepsilon)/2\Delta}.$$

The proofs. The lower bound for the Janson Inequality is immediate as

$$\Pr[\wedge \bar{A}_i] = \prod_{i=1}^{m} \Pr[\bar{A}_i | \bar{A}_1 \cdots \bar{A}_{i+1}] \ge \prod_{i=1}^{m} \Pr[\bar{A}_i]$$

by the correlation assumption and this equals M. The upper bound uses an *upper* bound on $\Pr[\bar{A}_i | \bar{A}_1 \cdots \bar{A}_{i-1}]$. Renumber so that i is adjacent to $1, \cdots, d$ and not to $d + 1, \cdots, i - 1$ in the dependency graph. Then

$$\Pr[A_i | \bar{A}_i \cdots \bar{A}_{i-1}] = \frac{\Pr[A_i \bar{A}_1 \cdots \bar{A}_d | \bar{A}_{d+1} \cdots \bar{A}_{i-1}]}{\Pr[\bar{A}_1 \cdots \bar{A}_d | \bar{A}_{d+1} \cdots \bar{A}_{i-1}]}$$

$$\ge \Pr[A_i \bar{A}_1 \cdots \bar{A}_d | \bar{A}_{d+1} \cdots \bar{A}_{i-1}].$$

By Inclusion-Exclusion, holding the conditioning fixed, this is at least

$$\Pr[A_i | \bar{A}_{d+1} \cdots \bar{A}_{i-1}] - \sum_{k=1}^{d} \Pr[A_i \wedge A_k | \bar{A}_{d+1} \cdots \bar{A}_{i-1}].$$

The first term is precisely $\Pr[A_i]$ by the properties of the dependency graph. The correlation assumption gives that each addend is at most $\Pr[A_i \wedge A_k]$. So

$$\Pr[A_i | \bar{A}_1 \cdots \bar{A}_{i-1}] \ge \Pr[A_i] - \sum_{i=1}^{d} \Pr[A_i \wedge A_j].$$

Reversing

$$\Pr[\bar{A}_i | \bar{A}_1 \cdots \bar{A}_{i-1}] \le \Pr[\bar{A}_i] + \sum_{i=1}^{d} \Pr[A_i \wedge A_j].$$

As $\Pr[\bar{A}_i] \ge 1 - \varepsilon$ and $1 + x \le e^x$ for all positive x we find

$$\Pr[\bar{A}_i | \bar{A}_1 \cdots \bar{A}_{i-1}] \le \Pr[\bar{A}_i] \left(1 + \frac{1}{1-\varepsilon} \sum_{k=1}^{d} \Pr[A_i \wedge A_k]\right)$$

$$\le \Pr[\bar{A}_i] \exp\left(\frac{1}{1-\varepsilon} \sum_{k=1}^{d} \Pr[A_i \wedge A_k]\right).$$

Multiplying these bounds over $1 \le i \le n$ gives $\Pr[\bar{A}_1 \cdots \bar{A}_n]$ on the left. On the right the $\Pr[\bar{A}_i]$ multiply to M. In the exponent the term $\Pr[A_i \wedge A_k]$ appears exactly once for each edge $\{i, k\}$ in the dependency graph. Thus they add to $\Delta/2$ giving a factor of $e^{\Delta/[2(1-\varepsilon)]}$ as desired. □

Now we turn to the proof of the Extended Janson Inequality. The proof of this probability theorem is by the probabilistic method! We begin with the reformulation of the Janson Inequality

$$\Pr[\wedge \bar{A}_i] \le e^{-\mu} e^{\Delta/[2(1-\varepsilon)]}$$

which we rewrite as

$$-\ln[\Pr[\wedge \bar{A}_i]] \ge \sum \Pr[A_i] - \frac{1}{2(1-\varepsilon)} \sum_{i \sim j} \Pr[A_i \wedge A_j],$$

with $i \sim j$ meaning i, j are adjacent in the dependency graph. For any $I \subseteq \{1, \cdots, m\}$ the same result holds for the conjunction of the $\bar{A}_i$, $i \in I$:

$$-\ln \Pr[\wedge_{i \in I} \bar{A}_i] \ge \sum_{i \in I} \Pr[A_i] - \frac{1}{2(1-\varepsilon)} \sum_{i \sim j; i, j \in I} \Pr[A_i \wedge A_j].$$

Consider a random set $I \subseteq \{1, \cdots, m\}$ in which $\Pr[i \in I] = p$ (p will be determined below) and the events $i \in I$ being mutually independent. Now both sides above become random variables and we can take their expectations! For every i the addend $\Pr[A_i]$ appears in $\sum_{i \in I} \Pr[A_i]$ with probability p so it adds $p \Pr[A_i]$ to the expectation, giving a total of $p\mu$. But when $i \sim j$ the addend $\Pr[A_i \wedge A_j]$ appears only with probability p^2 in the second term, so its expectation is $p^2\Delta/[2(1-\varepsilon)]$. Together

$$E[-\ln \Pr[\wedge_{i \in I} \bar{A}_i]] \ge p\mu - \frac{1}{2(1-\varepsilon)} p^2 \Delta.$$

We set

$$p = \frac{\mu(1-\varepsilon)}{\Delta}$$

(less than one by our side assumption) to maximize this giving

$$E[-\ln \Pr[\wedge_{i \in I} \bar{A}_i]] \ge \frac{\mu^2(1-\varepsilon)}{2\Delta}.$$

Hence there is a specific set I with the value at least this high so that

$$\Pr[\wedge_{i \in I} \bar{A}_i] \le e^{-\mu^2(1-\varepsilon)/2\Delta}.$$

But the conjunction of the $\bar{A}_i$ over *all* i is even smaller than the conjunction of the $\bar{A}_i$ over a restricted set I, giving the result. □

To some extent I think of the Janson Inequality as an inclusion-exclusion, stopping at the second term. Sometimes the second term is bigger than the first. By taking fewer terms, a proportion p of terms, we drop the first term by a factor of p but the second by a factor of p^2. By judicious selection of p the second is smaller than the first but the first is still fairly big.

Random graphs. In our application the underlying probability space will be the random graph $G(n, p)$. The events A_α will all be of the form that $G(n, p)$ contains a particular set of edges E_α. The correlation assumptions are then an example of a far more general result called the *FKG inequalities.* We have a natural dependency graph by making A_α, A_β adjacent exactly when $E_\alpha \cap E_\beta \neq \emptyset$.

Let us parametrize $p = c/n$ and consider the property, call it TF, that G is triangle-free. Let A_{ijk} be the event that $\{i, j, k\}$ is a triangle in G. Then

$$TF = \wedge \bar{A}_{ijk},$$

the conjunction over all triples $\{i, j, k\}$. We calculate

$$M = (1 - p^3)^{\binom{n}{3}} \sim e^{-\mu}$$

with $\mu = \binom{n}{3}p^3 \sim c^3/6$. We bound Δ by noting that we only need consider terms of the form $A_{ijk} \wedge A_{ijl}$ as otherwise the edge sets do not overlap. There are $O(n^4)$ choices of such i, j, k, l. For each the event $A_{ijk} \wedge A_{ijl}$ is that a certain five edges (ij, ik, jk, il, jl) belong to $G(n, p)$, which occurs with probability p^5. Hence

$$\Delta = \sum \Pr[A_{ijk} \wedge A_{ijl}] = O(n^4 p^5).$$

With $p = c/n$ we have $\varepsilon = O(n^{-3}) = o(1)$ and $\Delta = o(1)$ so that the Janson Inequality gives an *asymptotic formula*

$$\Pr[TF] \sim M \sim e^{-c^3/6}.$$

This much could already be done with the Poisson Approximation methods of Lecture 3. But the Janson Inequalities allow us to proceed beyond $p = \Theta(1/n)$. The calculation $\Delta = o(1)$ had plenty of room. For any $p = o(n^{-4/5})$ we have $\Delta = o(1)$ and therefore an asymptotic formula $\Pr[TF] \sim M$. For example, if $p = \Theta((\ln n)^{1/3}/n)$ this yields that $G(n, p)$ has polynomially small probability of being triangle-free. Once p reaches $n^{-4/5}$ the value Δ becomes large and we no longer have an asymptotic formula. But as long as $p = o(n^{-1/2})$ we have $\Delta = O(n^4 p^5) = o(n^3 p^3) = o(\mu)$ and so we get the *logarithmically asymptotic* formula

$$\Pr[TF] = e^{-\mu(1+o(1))} = e^{-(n^3p^3/6)(1+o(1))}.$$

Once p reaches $n^{-1/2}$ we lose this formula. But now the Extended Janson Inequality comes into play. We have $\mu = \Theta(n^3 p^3)$ and $\Delta = \Theta(n^4 p^5)$ so for $p \gg n^{-1/2}$

$$\Pr[TF] < e^{-\Omega(\mu^2/\Delta)} = e^{-\Omega(n^2 p)}.$$

The Extended Janson Inequality gives, in general, only an upper bound. In this case, however, we note that Pr $[TF]$ is at least the probability that $G(n, p)$ has no edges whatsoever and so, for $n^{-1/2} \ll p \ll 1$

$$\Pr[TF] > (1-p)^{\binom{n}{2}} = e^{-\Omega(n^2 p)}.$$

With a bit more care, in fact, one can estimate Pr $[TF]$ up to a constant in the logarithm for all p. These methods do not work just for triangle-freeness. In a remarkable paper Andrzej Rucinski, Tomasz Łuczak and Svante Janson have examined the probability that $G(n, p)$ does not contain a copy of H, where H is any particular fixed graph, and they estimate this probability, up to a constant in the logarithm, for the entire range of p. Their paper was the first and still one of the most exciting applications of the Janson Inequality.

Chromatic number. Let us fix $p = \frac{1}{2}$ for definiteness and let $G \sim G(n, p)$. In 1988 Béla Bollobás found the following remarkable result.

THEOREM. *Almost surely the chromatic number*

$$\chi(G) \sim \frac{n}{2 \lg n}.$$

The upper bound is given in Lecture 7. The lower bound, perhaps surprisingly, will be based on a large deviation result for the clique number $\omega(G)$, following the notation of Lecture 7. As done there let k_0 be the first k where the expected number of k-cliques goes under one. Now set $k = k_0 - 4$. Let A_α, $1 \le \alpha \le \binom{n}{k}$, be the events that G contains the various possible k-cliques so that $\omega(G) < k$ is the event $\wedge \bar{A}_\alpha$. We want to use the Janson Inequality. We calculate

$$\mu = \sum \Pr[A_\alpha] = f(k) > n^{3+o(1)}$$

as in this range $f(k)/f(k+1) = n^{1+o(1)}$. Now Δ is the expected number of edge-intersecting k-cliques. This is basically the second moment calculation done in Lecture 7. There we noted that the major term for Δ comes from k-cliques intersecting at a single edge and

$$\Delta = \Theta(\mu^2 k^4 / n^2).$$

We apply the Extended Janson Inequality to give

$$\Pr[\omega(G) < k] = \Pr[\wedge \bar{A}_\alpha] < e^{-\Omega(\mu^2/\Delta)} = e^{-\Omega(n^2 k^{-4})} = e^{-n^{2-o(1)}}.$$

Note that the probability of G being empty is also of the form $e^{-n^{2+o(1)}}$ so that $\omega(G) < k$ has "nearly" the same probability as G being empty—though certainly a $o(1)$ in the hyperexponent can hide a lot!

Now Bollobás's result is relatively elementary. Pick $m = n/\ln^2 n$. Let $k_0 = k_0(m)$ be the first k so that the expected number of k-cliques in $G(m, .5)$ goes below one and set $k = k_0 - 4$. As the original $k_0(n) \sim 2 \lg n$ we have $k \sim 2 \lg m \sim 2 \lg n$. The probability of $G(m, .5)$ having clique number less than k is $e^{-m^{2-o(1)}} = e^{-n^{2-o(1)}}$. Actually we want independence number but with $p = .5$ a graph and its complement have the same distribution so that the same bound goes for $G(m, .5)$

having no independent set of size k. The original $G(n, .5)$ has "only" $\binom{n}{m} < 2^n = e^{n^{1+o(1)}}$ m-element subsets so almost surely *every* m-set has an independent set of size $\sim 2 \lg n$. Let's color the graph $G \sim G(n, .5)$. Pull out independent sets of size $\sim 2 \lg n$ until there are less than m points left. (Since every m set has such a subset it doesn't matter that the points remaining are unrandom!) Give each set a separate color, using $\sim(n - m)/(2 \lg n) \sim n/(2 \lg n)$ colors. When less than m points remain give each point its own color. Wasteful, but this uses only $m = o(n/\lg n)$ colors. Together we have colored G with $\sim n/(2 \lg n)$ colors.

Index